Biomed:

From the Student's Perspective

Carlos R. Villafañe, CBET, CET

Biomed: From the Student's Perspective

Carlos R. Villafañe, CBET, CET
www.biomedtechnicians.com
www.techniciansfriend.com

ISBN: 978-1-61539-663-4

Copyright © 2009 by Carlos R. Villafañe.
www.techniciansfriend.com
All Rights Reserved.
Printed in the United States.

Graphics: www.planetabinario.com

The names of companies, their logos and products mentioned in this publication are Trademarks ™ of their respective companies.

The mention of products, companies, organizations, brands or authorities in this publication do not indicate that the author endorses them, or that the companies, organizations, brands or authorities endorse the author.

The images used in this publication are courtesy of the equipment manufacturers and has been given the corresponding credit.

This publication may not be reproduced in whole or in part, stored or transmitted by any means (whether mechanical, electronic, photocopying) without prior written permission of the author.

Biomed:

From the Student's Perspective

Carlos R. Villafañe, CBET, CET

About the Author

Carlos Villafañe has been involved in the electronics field since 1987. During his career, he has worked as an electronics technician, biomedical technician, consultant, electronics teacher and has provided technical support to various independent service repair centers. Carlos has studied Electronics Engineering, Biomedical Engineering, Computers and Communications. In 2007 he completed his Bachelor's Degree in Technical Management.

Currently he holds several certifications:

- General Radiotelephone Operator's License (GROL) with a Ship Radar Endorsement License, issued by the Federal Communications Commission (FCC).
- General Maritime Distress and Safety System (GMDSS), issued by the FCC.
- Chief Examiner by the National Radio Examiners (NRE), provides test reviews and commercial exams for the FCC.
- A + Certified Professional (Computer Repair) by CompTIA
- Certified Electronics Technician (CET)" Wireless Communications ", "Radar Electronics" and "Biomedical Electronics", issued by Electronics Technicians Association (ETA).
- Certified Biomedical Equipment Technician (CBET) issued by the Association for the Advancement of Medical Instrumentation (AAMI/ICC).

Carlos has collaborated with various specialized technical magazines in Spanish, including "Electrónica y Servicio" (Electronics and Service) in Mexico and "Saber Electrónica" (Knowing Electronics) in Argentina, where his technical articles have been published in the past. He has worked for Panasonic Sales Company, Vision Industries and Alpha Electronics, giving technical seminars at independent service repair centers in Puerto Rico and providing technical support to electronics engineering students at various universities.

The author worked as a Biomedical Technician for General Electric Medical Systems (GE Healthcare) from 2003 until 2008, servicing a wide variety of medical equipment. He currently works for BayCare Health System in the Clinical Engineering Department.

The author is currently the webmaster for Bay Area Association of Medical Instrumentation (www.baami.org). He's also the creator and webmaster for www.techniciansfriend.com and www.biomedtechnicians.com, created to provide technical guidance and information to students from various fields such as electronics, biomed and similar industries.

Special Dedication

I dedicate this work to my parents, Lydia and Nickey, who throughout my life have given me their unconditional love and support. I thank God everyday for having them. And to my wife Démaris, thank you for being by my side and sharing your life with me. I love you.

Acknowledgments

I want to thank my friend and brother Julio D. Vargas from Alpha Electronics (www.alphaelectronics.net), Mr. Charles McCaffrey and Mrs. Cynthia Osborne of GE Healthcare and Walter Barrionuevo of BayCare Health System for giving me the opportunity to work with them for several years and for their unconditional support in my career. I also want to thank my first Biomed teacher, Mr. Joel Castrodad Sánchez, for directing me towards this fascinating field.

Special thanks to my sister Ixia Villafañe for the revision and editing of the Spanish version of this publication, and to Ms. Paula Carey for proofreading the English translation of this book. *Paula: You're my hero!*

Thanks to my fellow coworkers: Alberto Perez and Edgardo Vega, who have helped me so much in the Biomed field. Also thanks to Jesus Tangarife (www.planetabinario.com), Julie Kirst from 24X7 Magazine, Robert King from AAMI and Barbara Kram from DotMed News for acquainting so many people with my work through their publications and the Internet. Thank you all!

Introduction to the English version of this book

The original version of this book was published in Spanish in 2008. It is titled: *"Biomédica: Desde la Perspectiva del Estudiante"* (ISBN # 978-0-615-24158-6).

The idea of writing this book in Spanish originated when I was studying Biomedical Engineering in Puerto Rico years ago. At that time, there were no publications of this kind in my native language. All our technical books came from the United States and (by default) in English.

That was a dilemma for many students whose first language is Spanish. To learn about their future career, all of them needed to learn English first in order to understand their technical books.

I also noticed that most technical books are limited to *theory*. There are few textbooks that explain what the Biomedical field entails, with real-life examples of a Biomedical Technician's job.

I had these concerns for years. After working in the Biomed field for 5 years, I decided to organize the information I had gathered and add my personal experiences. I have tried to explain things in a basic way, so that a new student or someone completely outside of the Biomed field could understand.

After publishing the book in late 2008, I started promoting it with various Biomed Organizations and on the Internet. I received book orders from Argentina, Dominican Republic, Puerto Rico, Mexico, and even Spain!

When Clinical Engineering publications in the USA like AAMI, 24X7 Magazine, DotMed News and others wrote articles about my book, something unexpected began to happen. I started receiving a lot of email requests: *"Do you have your book in English? We want your book in English"!*

The first opportunity I had to present the book in public was at a Bay Area Association of Medical Instrumentation (BAAMI) meeting in Tampa, Florida. There were dozens of biomed technicians in the meeting and I received good feedback from them. All of them gave me the same suggestion: *"translate it into English"*.

So, here it is, translated and updated! Whether you are a student or a technician, I hope that you enjoy it and good luck in your future endeavors!

Carlos R. Villafañe, CBET, CET
June, 2009

Preface

No matter who we are or where we live, technology has undeniably affected every aspect of our daily lives. From household appliances, computers, video games, microwave ovens, LCD screens, "hybrid" automobiles, GPS, cellular phones, PDA's, Ipods and especially in the Medical field, it is an incredible influence in our daily lives.

Today, all aspects of health care (prevention, diagnosis and treatment) rely heavily on computerized equipment. It is not unusual to learn about a machine or a newly developed surgical device that changes the way of diagnosing or treating a medical problem.

Each one of us that "submerge" into the world of technology usually does it because we *love* technology. We love the new gadgets and electronic equipment! However, many of us want to go beyond learning how to use an appliance; we want

to know how it looks on the inside, *how it works*. In fact, many of us that work in technical fields today started by disassembling our electronic toys when we were kids to see what was inside! We dreamed of becoming engineers or technicians to learn how to repair or design electronic equipment. That led us to the study of electronics, biomed, computer or similar fields.

The purpose of this publication is to provide more *practical* knowledge to students, especially those who are interested in the world of Biomedical Engineering, also known as "Biomed", "Clinical Engineering" or "Electromedicine".

When we begin to explore this field, several questions arise. What kind of equipment am I going to repair? What tools will I need? Is it dangerous to work with medical equipment? Is there "job demand" in this field? How much money will I earn? Do I need to know higher math?

Many of the Institutions offering Biomedical or Electronics courses include in their curriculum difficult math classes (pre-calculus, calculus), Physics, Technical Drawing, etc. ... Unfortunately, these courses lack *practical* help. As students, we do not find clear answers to questions like these. If our current teacher is not working in a hospital or the instructors don't keep up-to-date with the changes in the Biomed field, students have limited access to such information.

We graduate from our educational institutions well prepared at a *theoretical* level, but when we go into the industry we face a world that requires specific skills... and experience. As in many professions, there is a "jargon" that identifies technicians and biomedical engineers and the students should learn it in advance.

With this publication I wish to guide the student into the Biomed world. I want you to know beforehand what you will find in this broad area, especially if you have plans to work in a hospital, both in Puerto Rico and the United States. Why?

Because I believe that a student who has a fairly clear idea of the Biomed field *before* starting to work, will be better prepared to face the challenges offered by the field. That's why this publication would like to speak about the Biomed field from the perspective or viewpoint of a student.

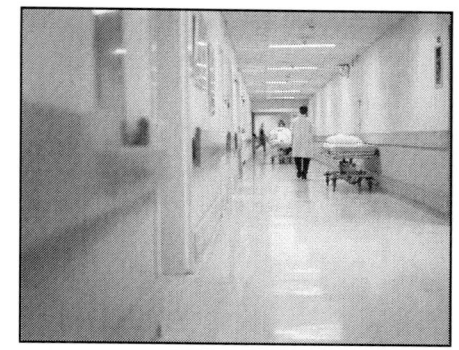

Hospital Hallway (Courtesy of Smith & Nephew).

It is worth mentioning that under no circumstances is this publication intended to be a comprehensive guide to an area as vast as Clinical Engineering. That would be impossible! There are so many different kinds of medical devices that I would need to create an entire encyclopedia with hundreds of volumes to mention all of them and it would need to be updated constantly.

With this publication I want to provide a simple guide for both teachers and students to help with the transition between school and industry and to explain how to progress in this field. Furthermore, I wish to present the information in a short, simple and understandable way for anyone who wants to learn about this fascinating field, without having to resort to deep theoretical data.

I hope that the information presented here will be helpful in achieving your future goals.

Good Luck!

Carlos R. Villafañe, CBET, CET

Contents

Introduction to the English version of this book

Preface

Part 1: What is a Biomed Technician?

Part 2: Differences between Electronics and Biomedical Technologies

Part 3: Electrical Safety
    3.1 What is an electrical hazard?
    3.2 Types of "Electric Shocks"
    3.3 Situations that could cause a macroshock
    3.4 Situations that could cause a microshock

Part 4: Some Biomedical Equipment
    4.1 Anesthesia Machines
    4.2 BP Machines
    4.3 Defibrillators
    4.4 EEG
    4.5 EKG
    4.6 Endoscopy
    4.7 Sterilizers (Autoclaves)
    4.8 ESU (Electrosurgery)
    4.9 Infusion Pumps
    4.10 Clinical Laboratory
    4.11 Patient Monitors

4.12 SpO2
4.13 Telemetry
4.14 Ultrasound
4.15 Ventilators
4.16 Radiographic Equipment and Digital Imaging

Part 5: Tools, Equipment Repair and the Biomed Department
    5.1 Basic Tools
    5.2 Physiological simulators and test equipment
        5.2.1 Safety Analyzers
        5.2.2 EKG Simulators
        5.2.3 Electrosurgery Testers (ESU)
        5.2.4 Defibrillators Testers
        5.2.5 NIBP Simulators
    5.3 The Clinical Engineering Department (Biomed Department)

Part 6: Regulatory Agencies, Accreditation and Legal Aspects
    6.1 JCAHO
    6.2 Medicare and Medicaid
    6.3 FDA
    6.4 OSHA
    6.5 AHCA
    6.6 NFPA
    6.7 UL
    6.8 NIST
    6.9 HIPAA

Part 7: Your Career as a Biomedical Technician
    7.1 Professional Certifications
        7.1.1 Do you need a Certification?
        7.1.2 CBET Certification
        7.1.3 CET-Biomedical Electronics

    7.2 Associations, Trade Shows and Biomed Conferences
        7.2.1 Networking
    7.3 Professional Publications and the Internet

Conclusion

Bibliography

Biomed: From the Student's Perspective

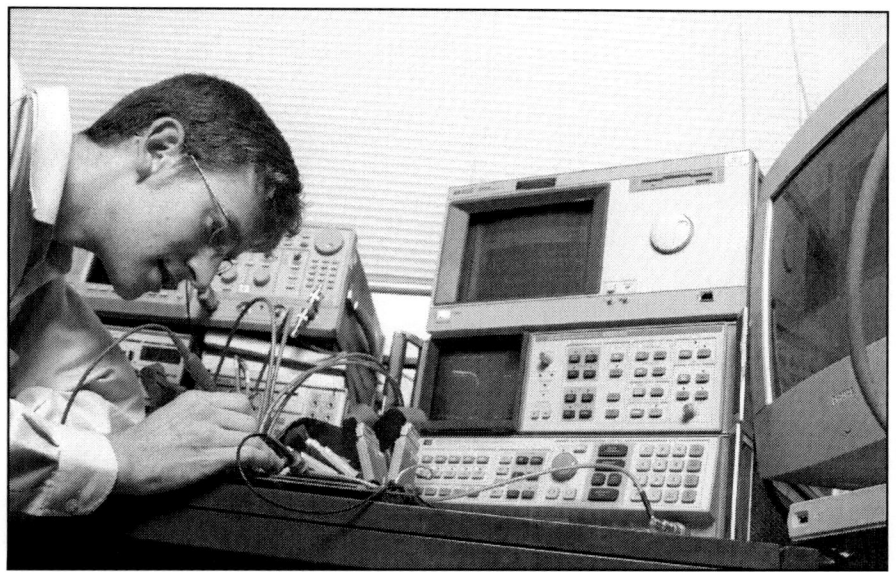

Courtesy of Philips

Part 1
What is a Biomed Technician?

What is a Biomed Technician? I've noticed as I have been working in this field that many people do not know what a *"Biomedical Technician"* is or what kind of equipment the biomed technician works with. A lot of people have never heard the term "Biomedical Technician" or "Clinical Engineer".

I've worked in various hospitals and there have been times when some employees of other departments have called us for *"collecting their biomedical waste"* *, when our job has <u>nothing to do with this!</u>

Figure 1 Biomedical Technician. (Courtesy of Philips).

Why did I start this publication telling you this? Because, despite the *crucial importance* of the Biomedical Technician's job in the healthcare field, unfortunately there are still many professionals who are unaware of the Biomed or Clinical Engineering department's roll in a hospital environment. Many are not even aware that *their work* <u>*depends*</u> on the care and preventive maintenance provided by the Biomed Department to their equipment in their own institution!

(* Note: Biomedical Waste can includes sharps (blades, needles, syringes and so on), liquid waste, waste crop strains, vaccines and mechanisms to inoculate or mix crops, animal waste (and parts of bodies animals), blood and its derivatives, tissue, etc.) [1]

Personally I have noticed that there is a lack of information about the Biomed field in the general public. I think it is the responsibility of each of us working in this area to publicize the value and the importance of our work. This has been one of the main reasons for publishing this book for Biomed students.

It is interesting that we can find several definitions of "Biomedical Technology", "Biomedical Engineering," or "Clinical Engineering." However, I believe one of the most comprehensive definitions is provided by the Biomedical Engineering Society (BMES):

"A Biomedical Engineer uses traditional engineering expertise to analyze and solve problems in biology and medicine, providing an overall enhancement of health care. Students choose the biomedical engineering field to be of service to people, to partake of the excitement of working with living systems, and to apply advanced technology to the complex problems of medical care. The biomedical engineer works with other health care professionals including physicians, nurses, therapists and technicians. Biomedical engineers may be called upon in a wide range of capacities: to design instruments, devices, and software, to bring together knowledge from many technical sources to develop new procedures, or to conduct research needed to solve clinical problems". [2]

Although this is the definition of a Biomedical Engineer, (someone that designs medical equipment and processes), the Biomedical Technician's responsibilities are very similar, except that they do not design biomedical equipment. Biomed Technicians are responsible for the repair, calibration and preventive maintenance of medical devices.

In a hospital, the Biomedical Technician is responsible for maintaining all medical electronic equipment in excellent condition in a timely, safe and responsible way for patient use and operation by the hospital staff. Biomedical technicians usually provide training to users (doctors, nurses, etc.) on the operation of the equipment and safety issues, thus providing safe and effective health care service.

However, the Biomed field is extremely broad and continues to expand as technology continues to advance. What skills are needed to become a biomedical technician?

A Biomedical Technician ("BMET" or Biomedical Engineering Technician) has the same expertise as an Electronics Technician. In many universities and colleges, if you compare a course handbook for both technologies you will notice that the "technical" classes that are offered in both courses are basically the same.

Figure 2　The Author repairing Biomedical Equipment.

Both technologies teach basic theory in terms of electricity and its components: Ohm's Law (the relationship between voltage, current and resistance). Both technologies explain the theory and operation of electrical circuits (series, parallel, series-parallel). The semiconductor component (diodes, capacitors, inductors, transformers, operational amplifiers, integrated circuits, etc.) theory is explained in both courses. Electrical safety concepts are explained and recommendations for diagnosing faults and failures, known as troubleshooting are taught too. Both courses teach schematic diagrams. Students are taught to use multimeters and other measuring devices. Students learn to solder and replace defective components.

So now the question is: Will it be the same thing to study Electronics Technology than studying Biomedical Technology? What differentiates one field from the other? Is one field more important than the other? Is an Electronics Technician prepared to work in the health care field? These questions will be answered in the next section.

Part 2:
Differences between Electronics and Biomedical Technologies

Electronics Technology and Biomedical Technology seem so similar. What are the differences between the two fields?

In my opinion, in simple terms and using a single word, the difference is: RESPONSIBILITY.

Why do I say this? Because the Biomedical Technician works with devices that are used in the care and treatment of patients. The Biomed working in a hospital environment is responsible for repairing and maintaining equipment designed to help keep other human beings alive. The health and life of another human being may be at stake! The priority of the Biomedical Technician is the safety of the patient and the user of the equipment.

On the other hand, the Electronics Technician typically works with electronic household devices such as televisions, radios, amplifiers, DVD's, CD players and so on. Others work on more sophisticated equipment such as communications equipment (cell phones, etc.) Even others specialize in computer systems, networks and similar equipment.

When repairing any of these "consumer electronics", "daily use" or "office electronics" there is flexibility in the repair of these units because there are many alternatives for

Figure 3 Common electronics parts. They are very easy to acquire.

replacement parts. Normally, the electronics technician works only with the electronic equipment to be repaired.

The Biomed Technician's job description is a little different. To perform the job effectively, a Biomed technician must learn not only about electronic repairs, but also:

- medical terminology
- human physiology
- electrical safety
- government regulations

An important factor for any Biomed tech is learning to work in a team. Teamwork includes dealing with people who do not work in the same field. The Biomedical technicians not only have to work directly with the repair and maintenance of medical equipment, but they have to be constantly involved with the doctors, nurses, Information Systems Department (I.S.) and hospital Administration to meet their requirements and needs. There are many government and regulatory agencies that work with hospitals. I will mention more details on these regulatory agencies in Part 6 of this publication.

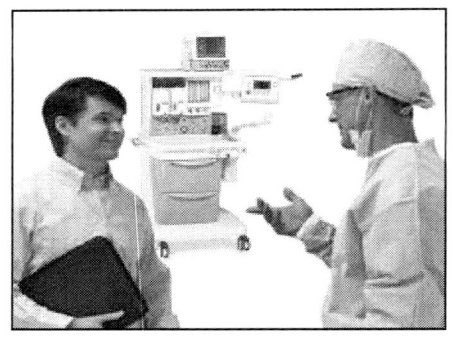

Figure 4 Good communication skills between the Biomed Technician and the hospital personnel is necessary. (Original Picture courtesy of Stryker)

Another appreciable detail worth mentioning is the integration between medical equipment and computer systems that has been happening in recent years. It does not matter whether you are studying Electronics or Biomedical Electronics, it is essential to acquire computers skills, knowledge of networks and the Internet.

For example, to calibrate and repair many medical devices it is necessary to use a laptop computer or custom computer programs. Many medical devices use operating systems very similar to those we use in our personal computers. The technology known as "DI" or "Digital Imaging" (which includes a computerized tomography (CT Scan) and magnetic resonance imaging (MRI), are completely dependent on computers!! Most of the

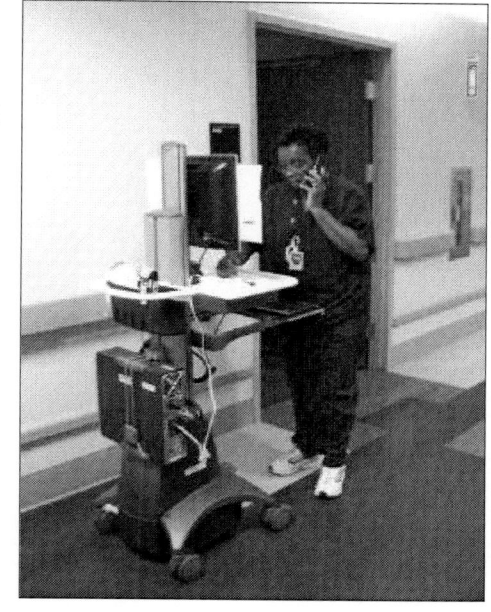

Figure 5 Modern "Computer on Wheels" System.

recent medical equipment has the ability to export information to the electronic medical record of the patient.

Due to these new features, many modern hospitals are changing to the "All Digital" concept, where all electronic devices are interconnected via networks. The results of the medical tests go directly from the machine that records the measurements or tests directly to the patient's medical record! What advantage does this provide?

When the system is well implemented, patient information is immediately accessible anywhere in the hospital. For example, imagine a patient who suffers an accident and goes to the emergency room (ER). When the patient's personal information is entered into the system, the data is readily available to the Radiology Department, which is physically in another area of the hospital. Perhaps once the patient receives an X-ray study, it is determined that surgical intervention is necessary. While the patient is transported from Radiology to the operating room, his health information and the tests results are already available in the operating room to be used by the surgeon!

Figure 6 Pepin Heart Center, Tampa, FL
"All Digital Hospital"

In summary, in the medical field today, computers are used in many areas: diagnosis, treatments, to control and maintain human life and to store and access patient medical records. My advise is: never stop learning about computers! Without a doubt, you will need computer knowledge at some point to perform your job!

The next section will analyze the main responsibility of a Biomedical Technician: ensuring that the equipment does not present electrical hazards to patients or users of the medical devices.

Part 3
Electrical Safety

One of the primary concerns in the Biomed field is electrical safety. Electrical safety is defined as "to contain and limit the damage caused by electric shock, explosion, fire or damage to equipment or the building" [21]. We must seek ways of limiting electrical risks in areas where services are provided to patients. Why? I will give you an example.

Imagine a patient who is in the middle of a surgery. Many of you have seen (perhaps in a television program) the vast amount of electronic equipment that is used during a surgical procedure: anesthesia machines, ventilators, "electrocautery"(ESU), Electrocardiogram machines (EKG), defibrillators to name a few. Figure 7 gives you an idea of the amount of equipment that can be found in an operating room.

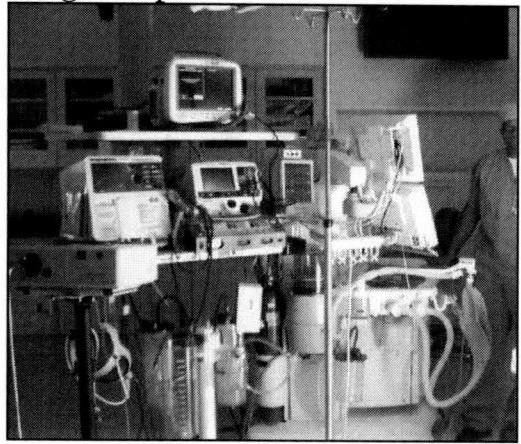

Figure 7 Common configuration of a modern surgery suite (Operating Room). Please note the extensive amount of electronic equipment.

It is known that technological advances in the medical field include risks, especially if the user of the equipment is not cautious.

Because surgery is an invasive * procedure, there are many electrical hazards that could fatally harm the patient if you are not careful.

An interesting fact is that Biomedical Technology, as a part of the healthcare industry, expanded rapidly in the 1970's. It was largely due to incidents that involved electrical hazards. An activist named Ralph Nader said that at that time: "At least 1,200 Americans die annually electrocuted during routine medical diagnostic procedures in hospitals." [3]

As more people became aware of this important issue, efforts to solve this problem gave way to the development of the several regulatory agencies in place today. These agencies specialize in preventing electrical hazards and other dangers in hospitals and Patient Care facilities.

Let's analyze some of these electrical hazards and determine what measures we can take to minimize them.

* (Invasive Procedure - It is a procedure, test or surgery that involves passing through the skin or muscles or into a vein or artery, such as cardiac catheterization [4])

3.1 What is an electrical hazard?

In Biomed, an electrical risk could be defined as a "dangerous condition created by a contact or failure of electrical or electronic equipment which could result in an electric shock or burns from an electric arc, temperature or explosion." Why explosion? Remember that in an operating room a variety of gases are used (oxygen, nitrous oxide, etc.). Years ago, some of the gases used were flammable,* such as: chloroform, ether, cyclopropane, etc. **[24]**. At that time, a short-circuit could cause a fire!

We must also bear in mind that electronic equipment produces "leakage current". Although it may be very small, in certain situations it could be fatal. It can cause an "electric shock" or a spark or a fire. What is "leakage current"?

Leakage current is defined as a *"current flowing through an alternate path, instead of traveling through the designated path when leaving or returning to the power source"*. **[5]** In other words, *"current flowing through an unwanted path"* **[6]** (i.e. the chassis of a computer).

Leakage current may be due to various factors, including:
• An accidental contact ("short") because of a problem in an electronic circuit or a component. It can also be caused by a blow received by the unit which allows the printed circuit or an internal component to touch the chassis.
• Elimination of the path to earth (ground) in the unit.
• Problems of capacitance or inductance in electrical equipment.

* (Flammable: Material that can ignite and burn rapidly. [26])

Exposed electrical wiring, broken insulation and similar problems are some causes of electrical hazards. All these factors mentioned above, make it necessary to check the physical integrity of medical equipment.

Feeling an electric shock is not pleasant and many of us have felt its effects at some point. Many of us have "respect" for electricity, because we know that it can be very dangerous to us.

iStockPhoto.com

For electricity to harm a person, the human body must become part of the electrical circuit. There are several important factors when considering the dangers of electricity, such as:

- The amount of current to which the person is exposed
- The path or direction of current passing through a person
- Time of exposure to electricity
- Weight and complexion of the person
- The current density (or concentration)
    - Major density on smaller current contact area
    - Minor density on larger current contact area

The severity of the injury will also depend on the organ(s) that is (are) affected by the shock.

I will briefly explain a number of additional details about how the human body is affected by the passage of electricity through it:

It is estimated that the dry human skin may have a resistance of approximately $15K\Omega/cm2$. The palms of the hands or the soles of the feet may have $1M\Omega/cm2$. Those resistive values are quite high, don't you think?

However, the problem is that the resistance of the skin can vary for many reasons, either by moisture (water, sweat, wet skin, or other liquids) or a laceration. Since the human body is composed of up to 78% water [7] and water conducts electricity, any open wound or laceration provides a lower resistive path to the flow of electricity. In some cases, the resistance would drop to less than $10\Omega/cm2$. Less resistance means more electrical current and exposes the patient to a condition where one could receive a power discharge or "shock".

Now let's see what types of "electric shocks" can affect us.

3.2 Types of Electric Shocks:

There are 2 main types of electric shocks: "Macroshock or Microshock.

A Macroshock occurs when a relatively high amount of electrical current flows through a human body. It can cause varying degrees of burns, wounds (the skin opens sometimes), cardiac fibrillation and death. Many define an electric shock as "electrocution."

In Figure 8, we see how a macroshock can occur:

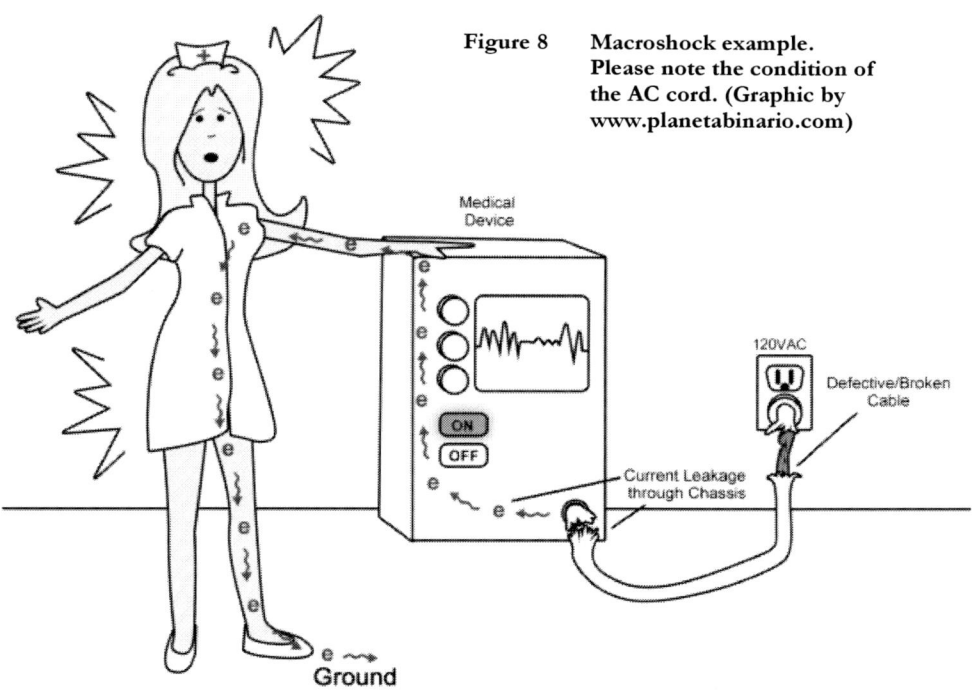

Figure 8   Macroshock example. Please note the condition of the AC cord. (Graphic by www.planetabinario.com)

Notice in the illustration that the power cord of the unit (AC cord) is torn /defective. Probably the contact to earth (ground) is broken. That causes the current "leakage" to take the path of least resistance to return to ground through the chassis of the device. In the illustration, the nurse becomes part of the circuit, providing the path to ground with little resistance and receiving an electric shock.

As we know, the human body operates on the basis of electro-chemical impulses. The current that normally occurs in the body is very low (in the range of microamperes µA). Let's see what the reaction of the human body is to various levels of electricity:

• Reference: Human, 155 pounds, medium complexion, dry hands, exposure of 1 to 3 seconds:

• 500µA or less = perception limit (do not feel anything)
• 1-5mA = tingling [24]
• 8-20mA = "let-go current" [24] (gives the feeling that you can't open your hands by involuntary contraction of muscles)
• 20-80mA = death by suffocation (lungs cease to function / respiratory paralysis) [24]
• 80-1000mA = cardiac fibrillation, internal burns, respiratory paralysis [24]
• 3A or more = cardiac paralysis and death

So, in short, electricity is one of the main dangers when using, maintaining and repairing all-electronic medical equipment.

However, we must know what causes a medical device to fail, and what could cause a Macroshock or Microshock. We will analyze this in the following sections.

3.3 What situations might cause a Macroshock?

As in the example in section 3.2, a very common situation that could cause a macroshock may be the physical integrity of a power cord (AC cord). Many times, these cables are broken or fail because people disconnect the unit by *pulling it by the cable*! What can happen? As we see in Figure 9, the internal wires can become exposed, creating a dangerous situation, not only for the patient but for any user of the equipment.

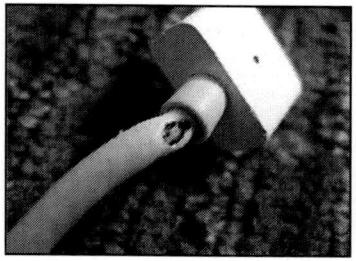

Figure 9 Cables exposed. Insulation is broken.

Another big mistake that occurs commonly is that some people eliminate the ground prong or "ground pin" on the plug. Removing that pin eliminates the built in protection of any electrical equipment! **NEVER** use a medical device on a patient if the physical integrity of the connector is compromised! If you find any device in the hospital that has the "ground pin" missing, (see Figure 10) ***take it out of service immediately*** * and replace it without delay with a new "hospital grade" connector (Fig. 11).

Figure 10 Broken "ground pin" in AC Connector.

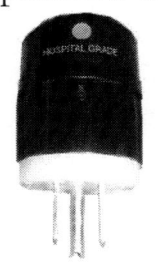

Figure 11 "Hospital Grade" Plug (Note the green dot)

(* Note: There's a Law requirement known as the "Safe Medical Devices Act (SMDA) 1990 makes it obligatory to inform or report any incident where the lives of a patient or a hospital team member is endangered. [8].)

Another scenario that often happens is that the electronic equipment gets wet. In an "environment of care," plenty of liquids are present: saline solution, liquid medicines, tube feeding, blood and other fluids. Accidents in which liquid contacts medical equipment are fairly common. Why?

Visitors and family members may bring food and drinks to the patient's room. They can bring a flower vase filled with water. A cup of coffee or a glass of water can cause havoc when it spills over an electronic device. Normally, a hospital team member uses liquid cleaning or disinfecting agents, and some spray the liquid <u>*directly onto the unit*</u>, rather than dampen a cloth with the cleaning solution.

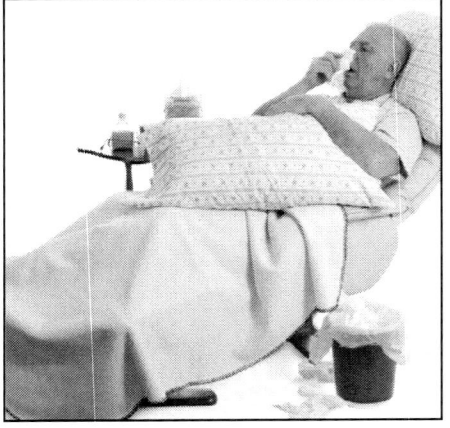

Figure 12 It's common that electronic equipment gets wet in a patient care environment. (iStockPhoto.com)

Any spill of fluid of any kind can create an electric risk. In addition, there are many portable electronic devices in use in the patient care area. Some are abused or might be dropped to the floor. Internal components of the device may make contact with the outer chassis, as mentioned earlier. That is why preventative maintenance of medical equipment is so important in the Biomed field!

Figure 13 Accidents involving liquids are common (iStockPhoto.com).

There is another type of risk that can cause an electric shock and could go almost *unnoticed*. It is known as "Microshock." Let's see why it is different than a "Macroshock" and why it is so dangerous.

3.4 What situations might cause a microshock?

A "Microshock" occurs when a small amount of current passes internally near the heart of the patient. How can the current flow enter the body and reach the patient's heart? There are procedures in which instruments are inserted into the body to investigate the condition of the veins and arteries leading to/from the heart. These procedures are known as "Catheterizations".

In the example I gave in the previous section (of a patient during a surgery), if only 250µA are applied directly to the heart (through a catheter), *it could kill the patient, without visible signs!*

To prove it, let's use Ohm's Law:

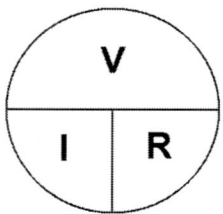

If the current (I) is 250µA and the Internal Resistance (R) is 10Ω:
V = I * R
V = 250µA * 10Ω = 2.5mV.

Figure 14 Ohm's Law.

With only 2.5mV, the electrochemical events that control the order of the muscle contractions of the heart can be affected, causing an immediate cardiac fibrillation. **[9]** Some sources indicate that only 0.1 mA directly applied to the myocardial tissue can induce a heart fibrillation **[10]**.

As mentioned earlier, a very small current known as the "leakage current" of any electronic appliance could cause problems for a patient. In simple words, in a patient-care environment, being careless during a procedure could cause a heart attack in a patient and possibly death!

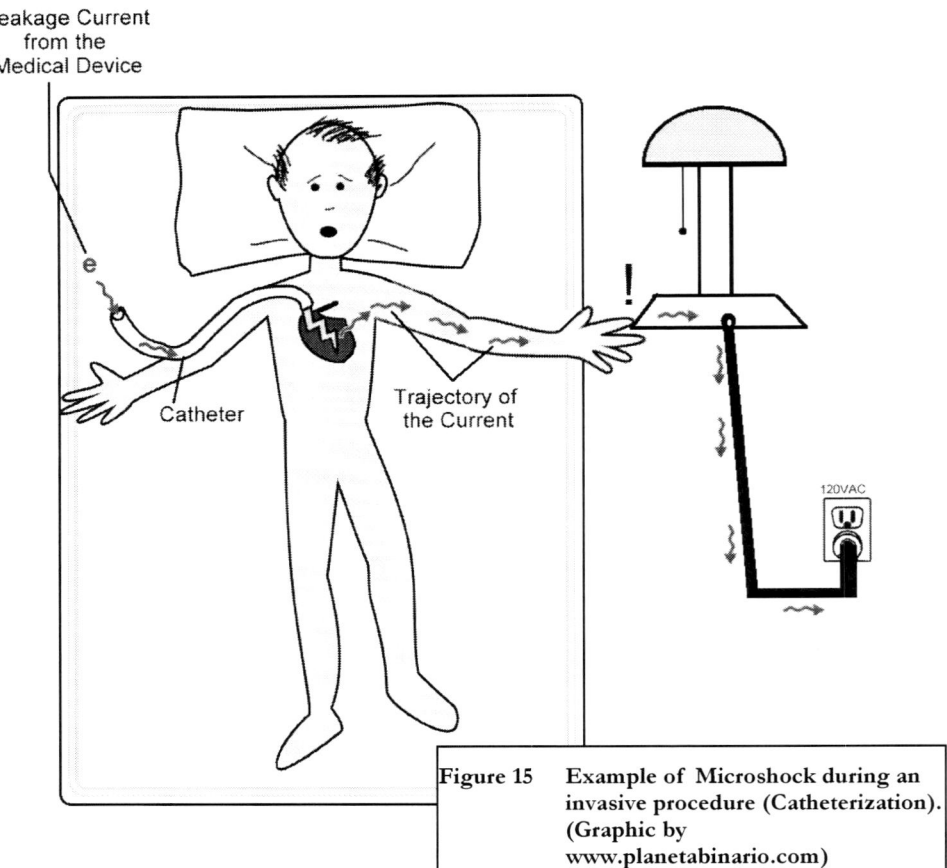

Figure 15 Example of Microshock during an invasive procedure (Catheterization). (Graphic by www.planetabinario.com)

In the hypothetical case shown in Figure 15, when the patient makes contact with the lamp, it is possible that both the equipment and the lamp were not "equally grounded". That causes the leakage current from the equipment to find a route to ground through the heart of the patient, possibly causing an immediate cardiac fibrillation.

It is therefore imperative that all electronic equipment that is used in any patient care area be tested physically and electrically, to ensure that it is safe to be used. As a result, partnerships have been created to develop "standards" or rules that govern every aspect of patient care, both in clinics and hospitals and other types of patient care facilities.

For example, AAMI (Association for the Advancement of Medical Instrumentation) is the regulatory agency dedicated to the safety and efficacy of medical equipment. AAMI is the primary agency responsible for guiding the Biomedical Technicians in the USA and offers the Biomedical Technician certification (CBET). There are more details on AAMI and the CBET certification in the next sections.

AAMI has specific codes regarding electrical safety. They include reports and best practices for preventive maintenance on all biomedical equipment. If you have Internet access, you can visit the AAMI home page for details: http://www.aami.org/standards.

Another regulatory agency that is involved in the electrical safety of biomedical equipment is the National Fire Protection Association (NFPA). Standard NFPA-99 indicates that the resistance to ground (chassis-to-ground) of any medical equipment *should be 0.5Ω or less.* **[11]** It is the duty of all the Biomed technicians to comply with these standards by performing a "Safety Test" or Electrical Safety Check to any medical equipment.

As you can see, complying with these rules is a major responsibility of the Biomedical technician. In addition, when you responsibly comply with these standards you are helping the hospital to provide quality service to all the patients.

As a future Biomedical Technician, **ALWAYS** remember this:

All the Biomedical Equipment that you repair, maintain, adjust or calibrate, can be used to save the life of a family member, the life of a friend, or your own life!

Part 4
Biomedical equipment

As you may have noticed, there is a large variety of medical equipment and various medical specialties. The devices you will repair in the future could depend on the specialty you choose. For example, if you decide to work in the Radiology and Digital Imaging field, you could work repairing X-ray machines, MRI's, CT scanners, or PET scanners.

Figure 16 Technicians repairing a "C-Arm". (Courtesy of Philips).

If you prefer laboratory equipment, you can repair hematology machines, centrifuges, urinalysis machines (Urine analyzers) and so on. If you select the general Biomed field, you could work with all sorts of equipment, from surgical equipment to digital thermometers. You have many alternatives to consider when you enter the Biomed field!

In this section you will see only a small sample of some well-known biomedical equipment that you will find in any hospital. I will not give you the theoretical explanations about the internal working of the equipment, because the theory will be explained in detail by your teacher in class.

I have included some photographs of actual units, so you can see how modern medical units look. The vast majority of the photos are devices that I have personally worked with in different hospitals. It would be very interesting for you, as a student, to have the opportunity of visiting a hospital bringing this publication with you. That way you can identify the equipment you see! That will help you become familiar with the devices and with the hospital environment in general. Try it!

Something that you will have the opportunity to observe is that, although there are a myriad of manufacturers and brands of biomedical equipment, many of the devices are quite similar, both in their physical appearance and their basic function. Some are designed for specific departments in the hospital (i.e. laboratory or surgery), while others can be found in any area of the institution.

Figure 17   Biomedical Technician (Courtesy of Philips).

4.1 Anesthesia Machines

The word "anesthesia" comes from a Greek root, "an" = "no" and "aestesia" = "feel." Therefore, it can be defined as "no feel" or "no pain". Anesthesia is commonly used before and during surgery, so the patient does not feel any pain or discomfort during the intervention.

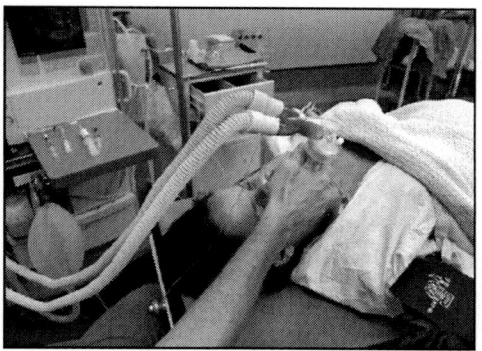

Figure 18 Supplying general anesthesia to a patient. (iStockPhoto.com).

Local or general anesthesia is achieved through drugs or other gases (nitrous oxide, Isofluorane, Desfluorane). In the case of general anesthesia, the surgery department has specialized machinery to administer the drugs at appropriate concentrations to the patient. Thus the patient is able to sleep in a safe manner.

As patients lose all muscular function during the procedure, the machine provides mechanical ventilation. Moreover, there is additional equipment to monitor the vital signs of the patient, the concentration of gases and other important details.

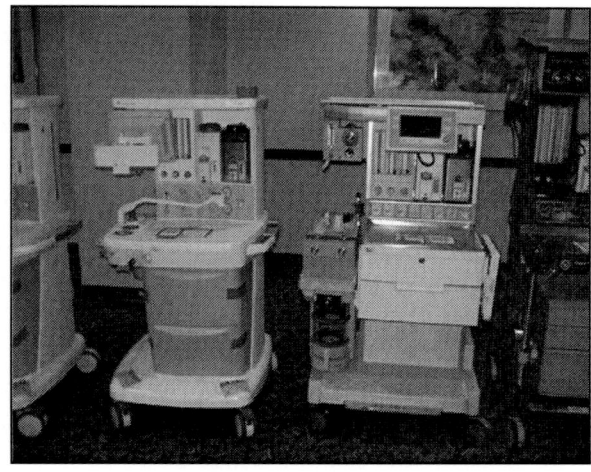

Figure 19 Examples of different brands and models of anesthesia machines.

The anesthesia machines have gas tanks on the back with the required gases or have hoses that connect to the air and gas systems of the hospitals. Connectors are clearly labeled on the walls or ceiling of the surgery suites.

Figure 20 Anesthesia machine basic controls. (iStockPhoto.com).

These devices have adjustable controls and valves that regulate the amount of air, oxygen, nitrogen oxide and other gases that are administered to the patient in the right amounts.

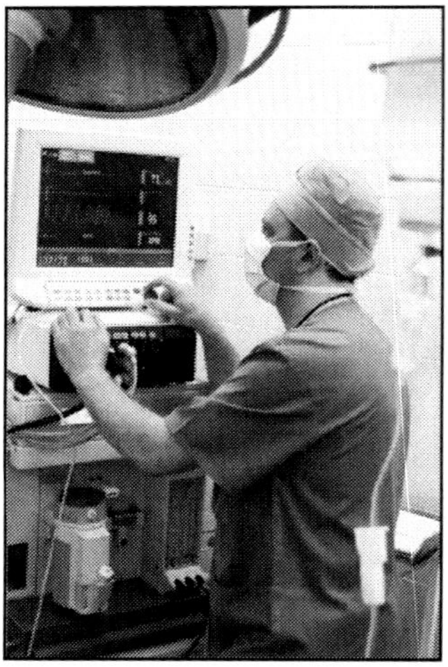

For each surgical procedure there is an anesthesiologist who has the responsibility of constantly monitor the patient's vital signs and the proper functioning of the equipment.

All anesthesia machines are considered "Life Support Equipment" or equipment used to sustain the life of patients. <u>You should not work on them without proper training</u>.

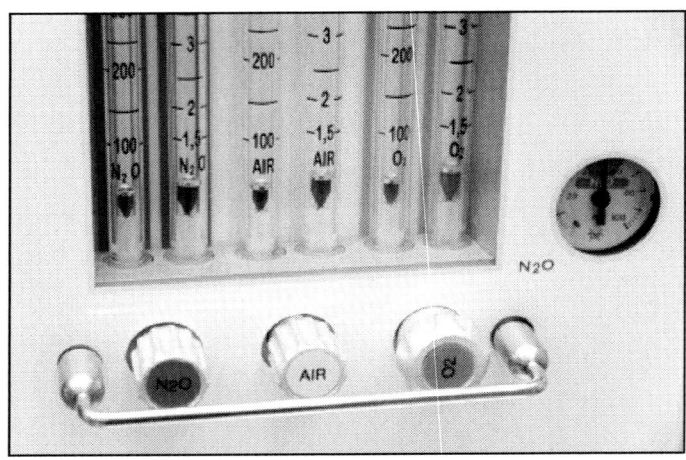

Figure 21 Anesthesilogist at work. (iStockPhoto.com).

4.2 BP Machines (Blood Pressure Machines)

Blood pressure machines exists in two forms: Invasive (which depend on a transducer that is placed internally in the artery of the patient) or non-invasive. In this section I will refer to non-invasive equipment, which is known as "NIBP's" - (Non-Invasive Blood Pressure).

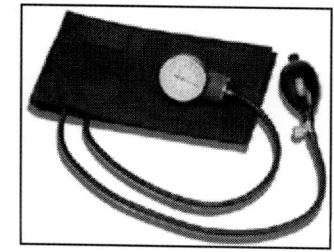

The correct name for these instruments is sphygmomanometers. The word comes from the Greek sphygmós (pulse), hands (not dense) and metron (measure). [12]

Figure 22 Basic manual Sphygmomanometer.

There are two types of equipment for measuring blood pressure, manual or mechanical (with a needle or display of mercury), and electronic (automatic). They are used to measure blood pressure by an indirect external compression of the

artery and adjacent tissues [13] for the measurement. This type of measurement is known as "noninvasive" as nothing enters the patient's body.

Now I will explain in simple terms how it works:

Figure 23 "Wall Mount" Sphygmomanometer. You can usually find it installed near the patient beds.

When the heart beats, it pumps blood through the body through your veins and arteries. The force of the pumping heart, the diameter of the veins and arteries and the blood volume that flows through them are the factors that determine blood pressure. When the heart contracts, the pressure increases. This is called systolic pressure. Between contractions, it is called diastolic pressure.

When you take the blood pressure values, you will notice that the result is given with 2 numbers. For example, you say "117 over 85." (117/85). The first value is the systolic pressure and the second value is the diastolic pressure.

Typical values or "normal" blood pressure measurements are:

- Systolic: 120 mmHg
- Diastolic: 80 mmHg

Depending on the values of this measurement the doctor or health professional will tell you if you have "high blood pressure" and the treatment you will need to normalize your blood pressure.

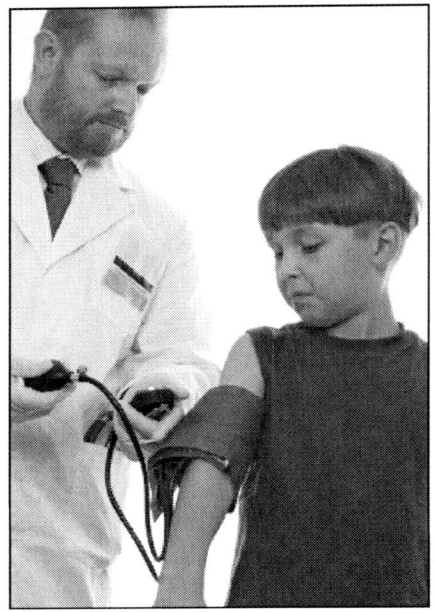

Figure 24 Doctor measuring blood pressure. (iStockPhoto.com).

How do you take a blood pressure? Normally, an inflatable cuff is placed around the patient's arm, at the height of the heart. Then the cuff is inflated with air (using a rubber bulb) until it stops the flow of blood through the artery in the arm.

Figure 25 shows the process of taking blood pressure with Sphygmomanometer.

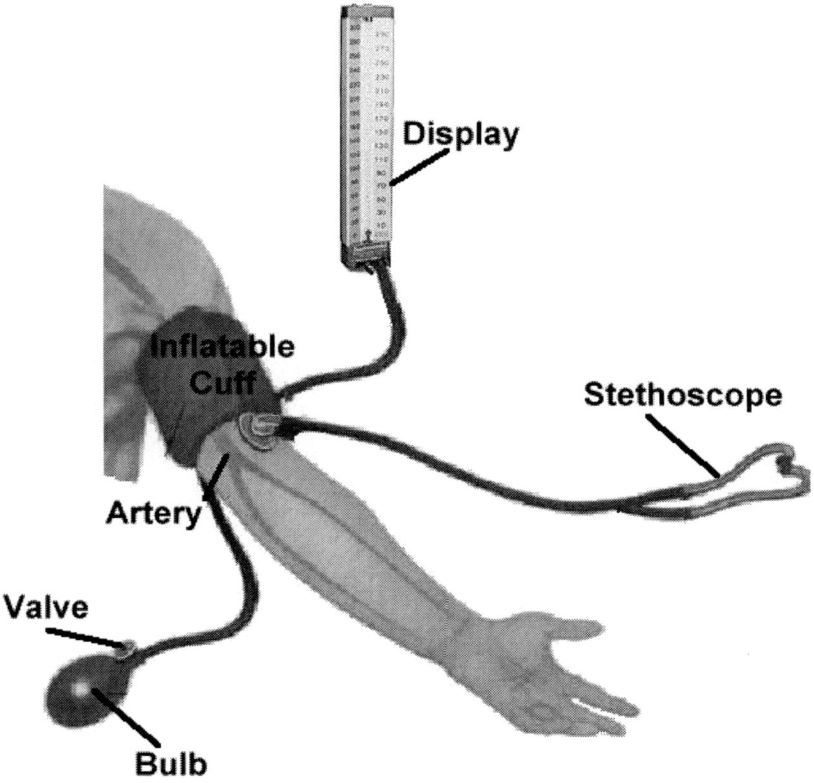

Figure 25 Using a Sphygmomanometer.

After inflating the cuff and interrupting the blood flow, the doctor monitors the pressure in the display. Then he begins to let the air out slowly through a small valve on the cuff. When blood begins to flow again, "oscillations" start to occur *.

* (Note: This oscillations are known as "Korotkoff sounds" for a Russian surgeon called Nikolai Korotkoff, who discovered them and developed the auscultation method)

These variations can be heard through a stethoscope, a microphone, or a sensor (transducer). When the oscillation starts it indicates the systolic pressure. The moment that you no longer hear the oscillations is known as diastolic pressure.

Figure 26 shows some of the modern electronic devices that are used to measure blood pressure. These devices operate on the same principle of "oscillometry". The electronic system performs the whole process automatically and gives you the measurement results in a digital display. Some models include a printer and you can print the results.

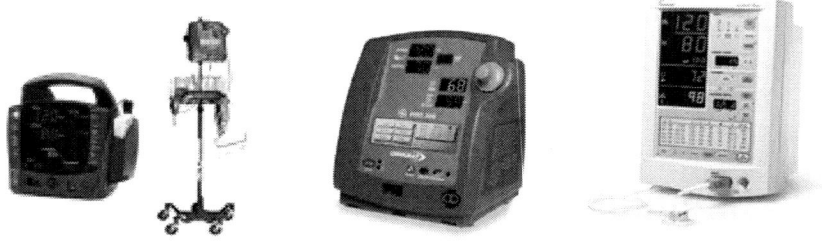

Figure 26 Blood Pressure Machines. Courtesy of GE Heathcare and Datascope.

4.3 Defibrillators

A defibrillator is the device that is used to stop a heart fibrillation (what could result in a "heart attack"). In fact, a fibrillation is defined as an "uncoordinated contraction of the heart".

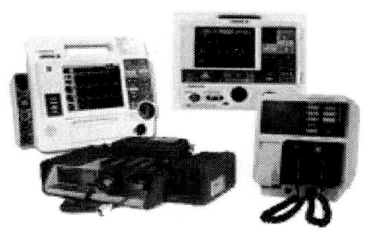

Figure 27 Defibrilators. (Courtesy of Medtronic Physio-Control).

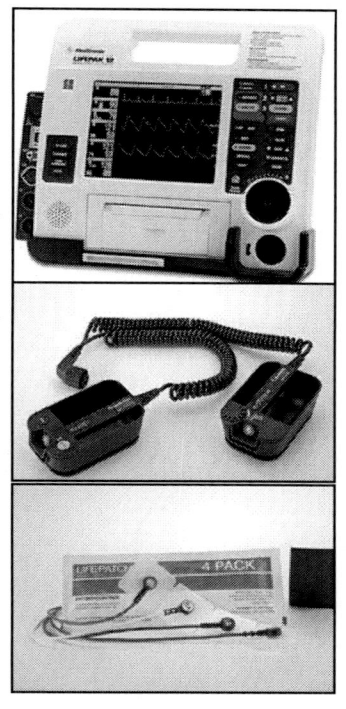

Figure 28 Defibrillator, Paddles and Electrodes. (Courtesy of Medtronic Physio-Control).

A cardiac fibrillation may be due to problems in the conductive system through the heart. The contractions of the heart muscle are "out of timing", you lose control over it and it starts beating wildly, to the point that it ceases to pump blood. At that time the patient no longer has heartbeats, a pulse and one can not detect any blood pressure [14].

The defibrillator has been developed and refined over decades and is the main tool used to save the lives of many people daily. The defibrillator is usually composed of a central unit with a screen and two paddles or electrodes emerging from the main unit. The paddles or electrodes make contact with the patient's chest. The unit has buttons that are pressed for "charging" the unit to deliver the energy to the patient.

There are electrodes that can be attached to the skin of the patient directly. This is useful for monitoring the vital signs and to deliver the charge to the patient as soon as necessary (without requiring the paddles).

A common defibrillator has internal capacitors that, when charged to the capacity selected, can produce up to 360 joules* of energy. By discharging its energy, the defibrillator produces a strong electric shock to the patient's chest from one electrode to the other to "depolarize" the heart's electrical system. This provides the heart with an opportunity to "reset" or start beating again, gaining control of the muscle tissue. [13] It is imperative that a defibrillator be available in all areas of the hospital.

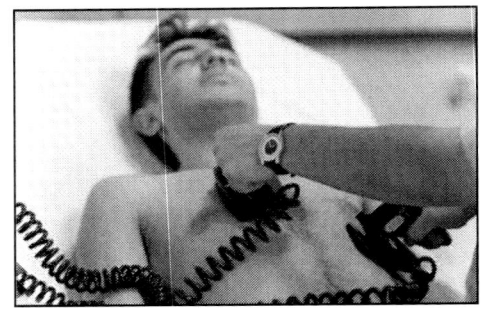

Figure 29 Using a Defibrillator. (Courtesy of iStockPhoto.com).

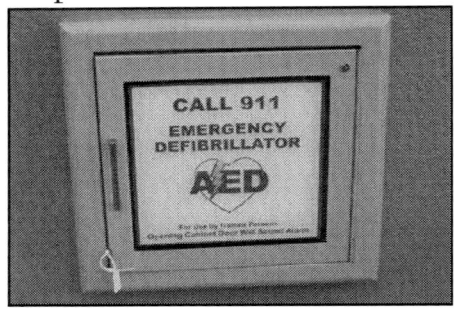

Figure 30 AED (Automatic External Defibrillator)

Today it is common to find portable units called AED (Automated External Defibrillator) wherever groups of people gather together, both public and private.

*Joules - Measurement of electric power. 1 Joule is equal to 1 Amp of current through a resistance of 1 Ohm, for a period of 1 second. [15]

All defibrillators are considered "Life Support Equipment" or equipment used to sustain the life of a patient.

This is another example of machines that need extra attention and concern when servicing. Doing the proper maintenance and checking the status of the batteries on each unit is of paramount importance.

4.3 Electroencephalogram (EEG)

It comes from "electro" (or electrical activity), "encephal" (brain) and "gram" (a Greek root meaning "to write"). So Electroencephalogram is a machine that records the electrical activity in the brain, amplifies it, processes it and displays it on a monitor or in printed form.

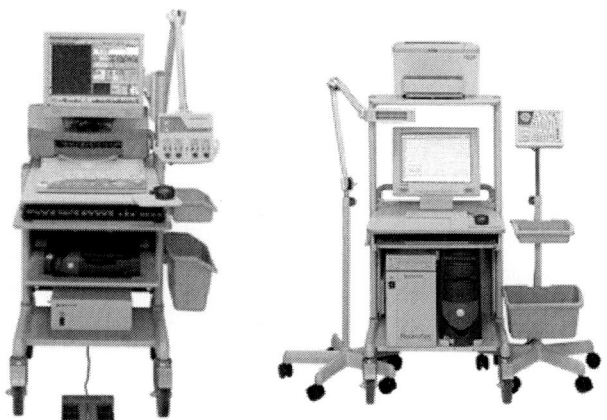

Figure 31   Electroencephalogram machines (EEG). (Courtesy of Nihon Kohden).

An Electroencephalogram is used to diagnose diseases and conditions that affect the brain, such as epilepsy (seizures), brain tumors, encephalitis and similar conditions. How does it work?

Brain cells transfer messages to each other by means of very small electrical currents. That makes it possible to measure them on the surface of the head.

An EEG machine looks like a normal computer system, except that it includes an external module with many cables of small electrodes. These small electrodes are placed on the patient's head at specific points and they capture those small brain signals.

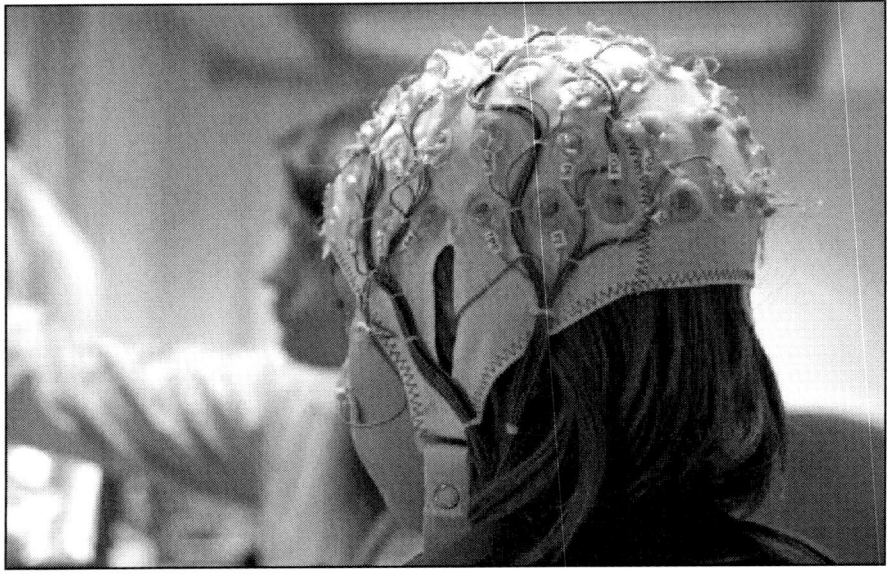

Figure 32 EEG procedure. Note the number of electrodes used in the test. (Courtesy of iStockphoto.com).

The computer system processes the signals immediately and the brain activity is shown on the screen or on paper as a series of lines that represent the electrical activity in different areas of the brain.

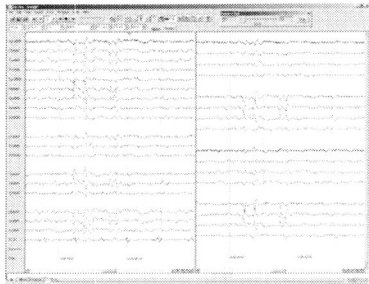

Figure 33 Brain activity as shown by a modern EEG system.

4.5 EKG or ECG - Electrocardiogram

The term "electrocardiogram" consists of "electro" (electrical activity), "cardio" (heart) and "gram" (a Greek root meaning "to write"). An electrocardiogram is a machine that records the electrical activity of the heart, amplifies it, processes it and displays it on a monitor and / or in printed form. It is a non-invasive method which can measure the current flow on the surface of the body through electrodes that are connected to the patient's chest and limbs.

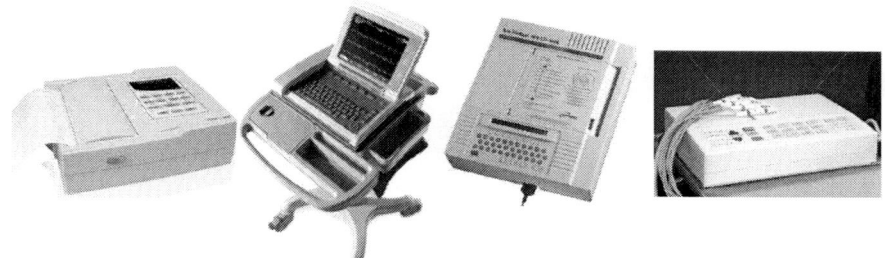

Figure 34   Various ECG Units from Bionet, GE/Marquette and Burdick. Last photo courtesy of iStockPhoto.com.

The heart has a "conductive" system, which uses electrical impulses to coordinate the contractions (heartbeats). An EKG machine is used to detect the electrical activity of the heart muscle (myocardium). By monitoring these signals, cardiac conditions can be identified: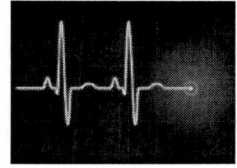
abnormal heartbeats (too slow or too fast), arrhythmias and so on. You can also identify if the heart muscle is damaged, if it is "pumping" with enough strength, birth defects and many other conditions.

The doctor can perform these tests while the patient is resting or while doing certain exercises. (Running on an exercise machine is known as a "stress test" – see Figure 35). There are portable EKG machines that record information for an extended period of time. They are known as "Holters.

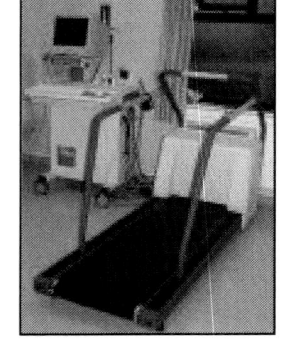

The image of the cardiac signal on the computer screen or on paper, is divided into small squares of 1 mm. 10 of these pictures represents (vertically) 1mV and horizontally, the paper moves at a speed of 25mm per second. So, a block of 1 cm square is 0.04 seconds (or 40ms) **[28]**.

Figure 35 "Stress Test Machine"

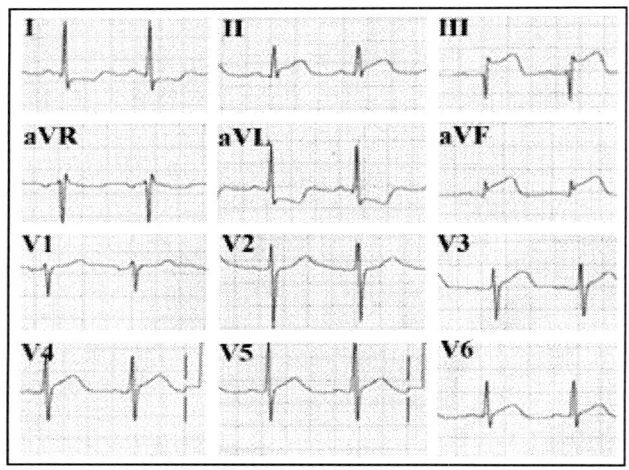

Figure 36   How an EKG signal looks printed on paper.

4.6 Endoscopy

"Endo" is a prefix meaning "internal" or "inside" **[16]**. In a hospital, the Endoscopy Department has specific tools to "look inside" the patient's body using its natural openings. (Therefore, it is *not* considered an invasive process).

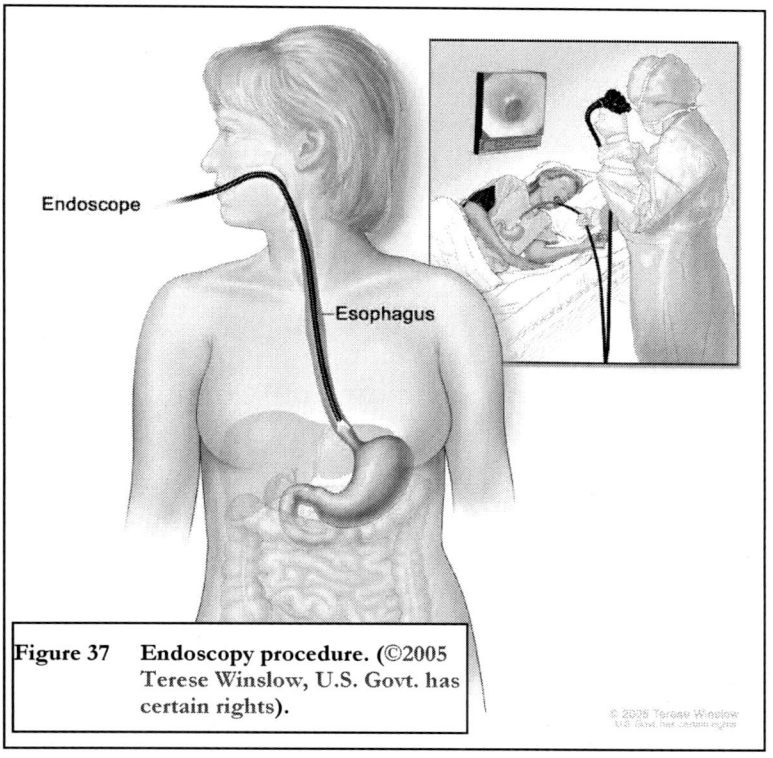

Figure 37 Endoscopy procedure. (©2005 Terese Winslow, U.S. Govt. has certain rights).

The name of the intervention and the name of the instrument are determined by the part of the body to be observed:

• Colonoscopy (colonoscope) = colon
• Duodenoscopy = duodenum (section of the small intestine that connects the stomach with the jejunum)
• Gastroscopy (gastroscope) = stomach, digestive system, etc.
• Bronchoscopy (bronchoscope) = lungs

The instrument to be used can be rigid or flexible, depending on the area you want to examine.

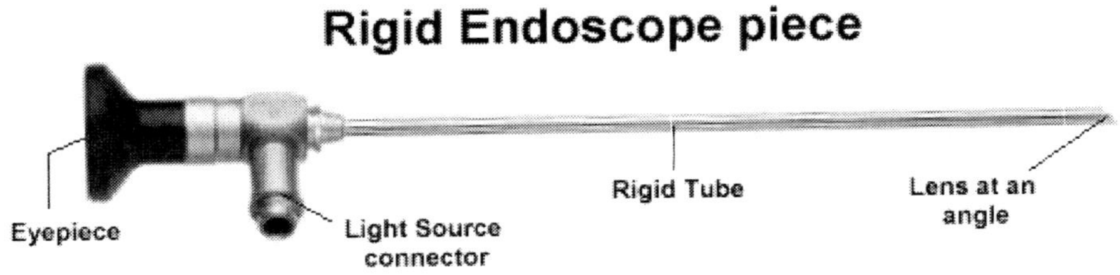

Rigid Endoscope piece

Figure 38   Rigid Endoscope

Some instruments are rigid, made of metal, and have a lens with a prism on the tip, which is calibrated at specific angles to observe the inside of the body. The instrument can be connected to an external light source to illuminate the area. The light is transmitted through a flexible fiber optic cable from the source to the instrument. Some instruments can be connected to an external video monitor as well as to a computer. That allows to keep a record of the intervention and to print photos for future analysis.

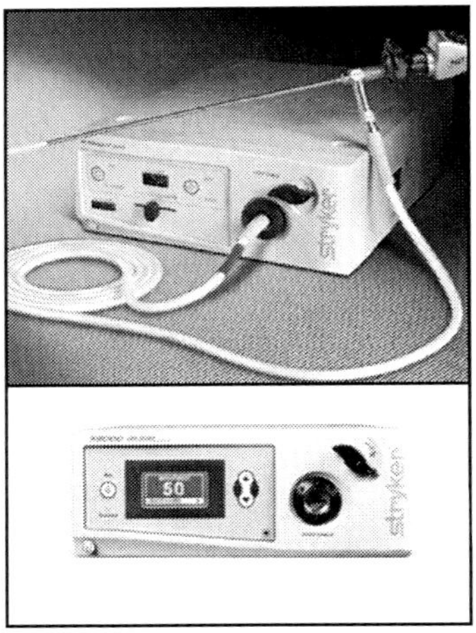

Figure 39   Light Sources. (Courtesy of Stryker)

On the other hand, a "flexible endoscope" is a long tube, very flexible and very thin, which has a tiny video camera at the tip. Some models have one or more internal channels through which the doctor can insert tools to take samples (biopsies). They can be used also to suction or pour water to the area under examination. They are also lit by fiber optics from an external light source.

Figure 40 shows how a flexible endoscope looks:

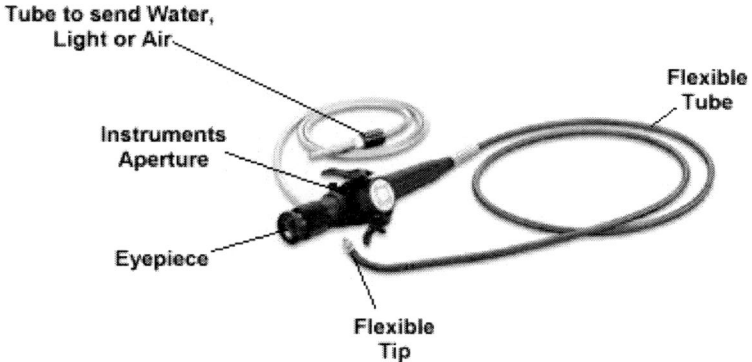

Figure 40 Flexible endoscope.

The surgeon or specialist can guide the instrument within the patient's body, because the endoscope has manual controls and a mechanism to bend the tip in all directions to precisely check the desired area.

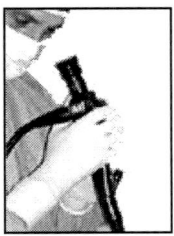

Figure 41 Using an Endoscope.

4.7 Sterilizers ("Autoclaves")

Sterilizations machines are used to eliminate *"all forms of life, including spores or microorganisms contained in an object or substance ... in a way that prevents further contamination."* [17] In other words, they are used to kill viruses and all kinds of transmissible agents such as fungi, bacteria, parasites, etc. from any surface.

Sterilization is very important to avoid infections. Any equipment or instrument used during surgery (such as scalpels, transducers, and so on) *must* be sterilized before each procedure, to protect both the patients and the surgeons.

The steam sterilizers operate much like a pressure cooker.

Figure 42 "Pressure Cooker".

Under normal conditions, the water boils at 100° C and it is not possible to increase the temperature more than that. However, in a sealed container such as a pressure cooker, the boiling point of the water increases because of the pressure created by the steam while expanding. Such high temperatures and pressure kills bacteria. (And it also cooks our food in the kitchen! ☺)

Figure 43 Sterilizer. (Courtesy of Steris Corporation).

An "autoclave" or Industrial Sterilizer physically looks like a big refrigerator on the outside, but internally it works the same way as the pressure cooker. Surgical instruments to be sterilized are inserted inside the unit and sealed hermetically. The unit has controls to regulate the temperature, pressure and exposure time. Most of them have a digital display and a printer.

There are other machines that are used to sterilize and do not necessarily depend on heat or pressure. Why? Obviously, not all materials are resistant to heat. Plastics and other materials can melt, degrade, break or discolor. So there are other machines that achieve sterilization using chemicals, gas or filters.

For example, as mentioned in the Endoscopy section, colonoscopes and other instruments are made of materials that are susceptible to heat (plastic, rubber) and others have lenses and prisms. There are specialized machines that sterilize instruments such as these, like the one showed in Figure 44.

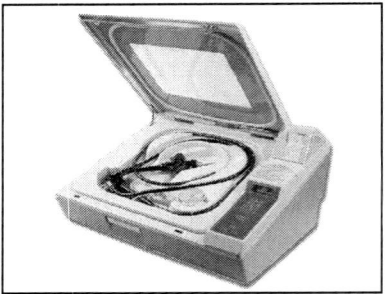

Figure 44 STERIS SYSTEM 1™ Sterile Processing System. (Courtesy of Steris Corporation).

4.8 ESU / Electrosurgery (Electro Surgical Units)

Electrosurgery machines are used to cut tissue and to stop bleeding or hemorrhage during surgery. This is accomplished through "blood clots". How does it work?

The machine consists of a central unit (generator) and it has an electrode (ground pad) that connects to the patient and a "pencil", which is used as a scalpel by the surgeon.

The generator produces a radio frequency (RF) which is adjusted using the controls in the central unit. That RF signal can cut the tissue or causes the blood to clot.

For example, if you set the output power around 500 KHz, the "pencil" cuts the skin and tissues [27]. The current that is generated between the "pencil" and the tissue evaporates the water in the cells, destroying them on contact. This not only opens the skin or tissue, but cauterizes it at the same time, avoiding bleeding.

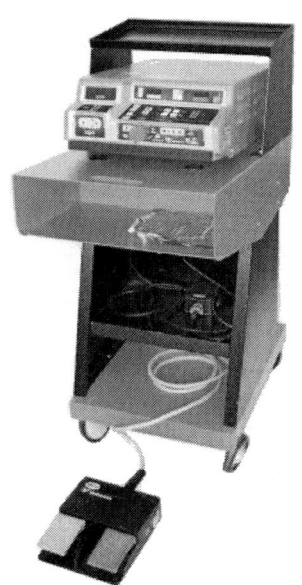

Figure 45   Electrosurgical Unit (ESU). (Courtesy of Valleylab/Covidien).

The process is controlled by the surgeon using pedal switches on the floor (see Figure 45). One pedal is used to CUT and the other is used for coagulation (COAG). Some machines have a feature known as "Blend", which combines the functions of coag and cut in one step.

The preventive maintenance and repair of this kind of equipment is extremely important. A device poorly repaired, misused or out of calibration can cause unwanted burns and cuts, both to the patient and to the surgeon. It can cause explosions as well, if used in the presence of flammable gases.

Figure 46 shows pictures of different Electrosurgery machines from three manufacturers

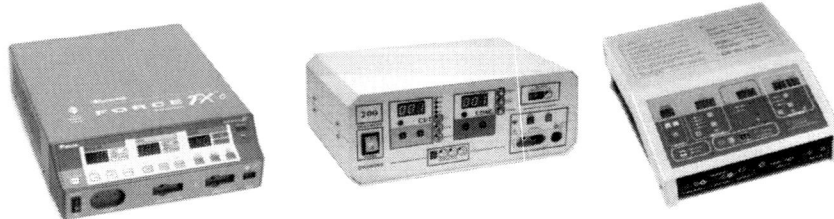

Figure 46   Various ESU Units.

 Note: As an interesting fact, the ESU machine was invented by a doctor named *William T. Bovie* while working at Harvard University in the 1920's. Now, most of the medical staff in the hospitals refers to the ESU machine simply as a *"Bovie"*.

4.9 Infusion Pumps (IV Pumps)

Infusion pumps are machines used to inject, infuse or administer liquids, medicines or nutrients to a patient's circulatory system (such as saline solution, dextrose solution, blood, medications, parenteral nutrition or others). These units are identified as "IV Pumps" (IV for Intra-venous).

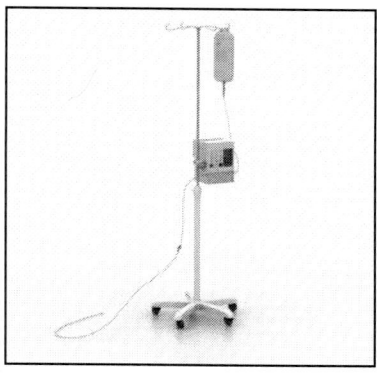

Figure 47   Infusion Pump. Picture shows a unit installed on a portable IV Pole.

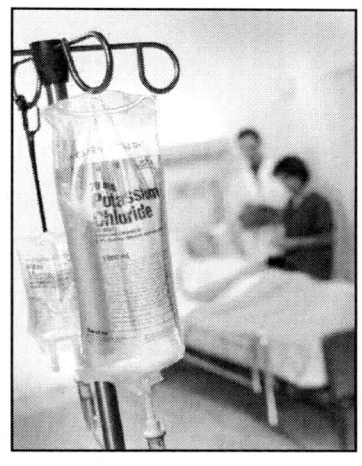

Figure 48   Solution bag on IV Pole. (Courtesy of Baxter).

These machines can be programmed to dispense the product to the patient automatically and they have controls to adjust the volume and speed (rate) of the infusion. All models have some security mechanisms to avoid over-infusing a patient and an alarm goes off if there is an obstruction or air in the tubing.

The product to be infused usually comes in a bag or a bottle that is hung on a pole above the unit. A tube goes from the bag to the patient's vein passing through the machine. The machine has a mechanism that squeezes the tubing at intervals *, infusing the product to the patient.

* This action is known as 'Peristaltic movement" since it resembles the way food moves through our gastrointestinal system.

Typically these machines are portable and they can be found in any department in the hospital. Figures 49 and 50 show how some of these devices look.

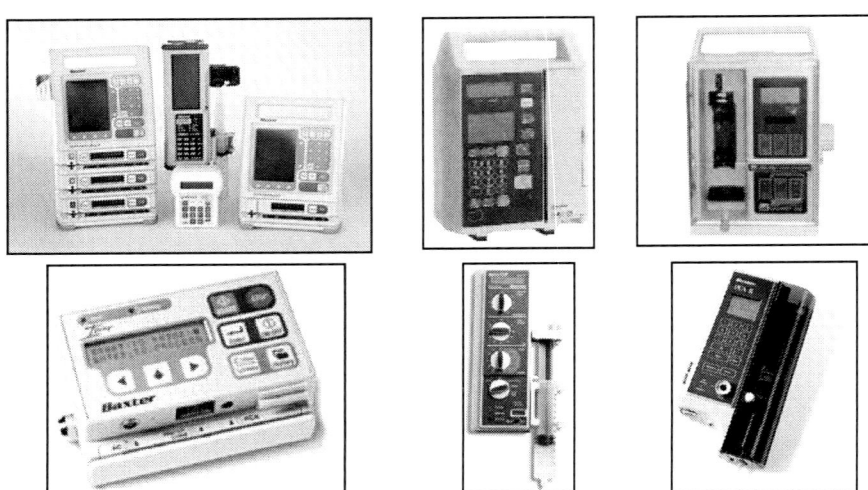

Figure 49   Infusion Pumps, Syringe pumps and PCA pumps. (Courtesy of Baxter).

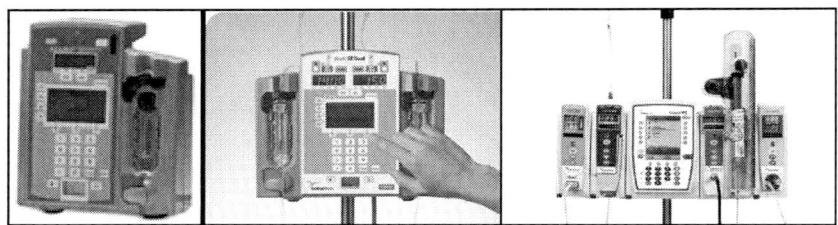

Figure 50   More Infusion Pumps. (Courtesy of Alaris/Cardinal Health).

4.10 Clinical Laboratory

Clinical laboratory testing plays a crucial role in the detection, diagnosis, and treatment of disease. The clinical laboratory is where *"up to 80% of the decisions that a doctor makes regarding diagnosis and treatment of a condition are based on laboratory results"* **[33]**.

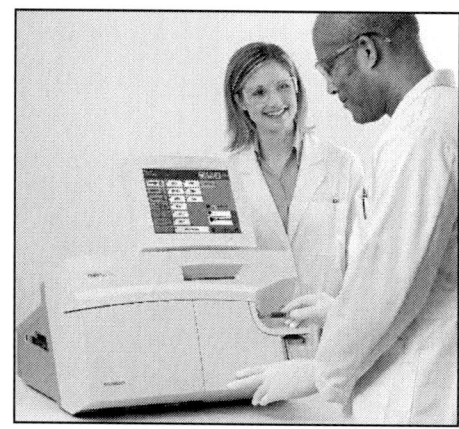

Figure 51　Laboratory technicians at work. (Courtesy of Siemens).

The Clinical Laboratory has evolved over the years to become a crucial department in all health centers. Technology has changed the way of performing various tests, because today we have faster, safer and more accurate machines.

There is a lot of different electronic equipment that is used today in a Clinical Laboratory. We will need another book just to mention the great diversity of units that exist.

There are devices for each specialty: hematology, pathology, microbiology, biochemistry, urinalysis and so on.

Figure 52　Modern Clinical Laboratory systems.

Most of the clinical devices are completely computerized with highly complex machinery.

Some devices have the ability to take samples (blood, urine) and analyze them with virtually *no human intervention*, using "conveyors" electronic arms and computer programs. The repair and calibration of these units is a very sensitive and accurate process.

Figure 52 Many units are completely automated.

If you decide to become a Biomed/Lab Technician, you will probably work with the following equipment:

• Centrifuges – These machines are used to separate substances of different densities. A motor rotates the sample at a high speed and centrifugal force takes care of the rest.

For example, if you put a test tube with a sample of blood in a centrifuge, at the end of the process you will see that the blood is physically separated into several components (plasma, etc).

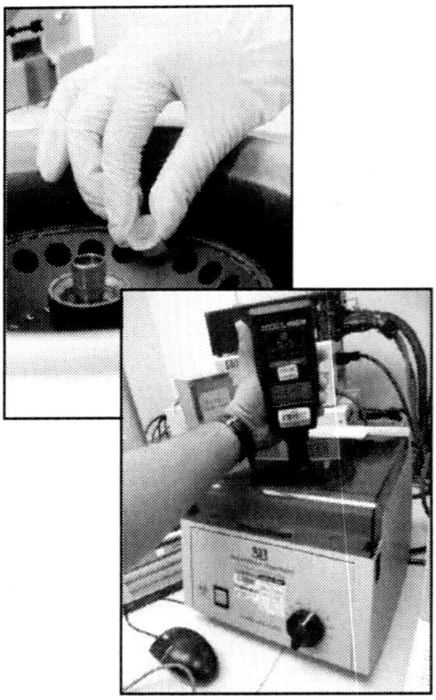

• Microscopes - There are many types of electronic microscopes with auto-focus systems and connectors for transferring data to a computer. The proper cleaning, adjustment and calibration of these units is very important.

Figure 53 Centrifuges. The Author is measuring RPM's in the second picture.

- Blood Analyzers (Hematology)

Blood analyzers are devices that are used for blood tests such as a "CBC" or Complete Blood Count, (which includes hematocrit, hemoglobin, red and white blood cells, etc.). It is one of the most sophisticated pieces of equipment that exists in clinical laboratories. Within seconds, the machine analyzes a small sample of blood by spectrophotometry and prints the results.

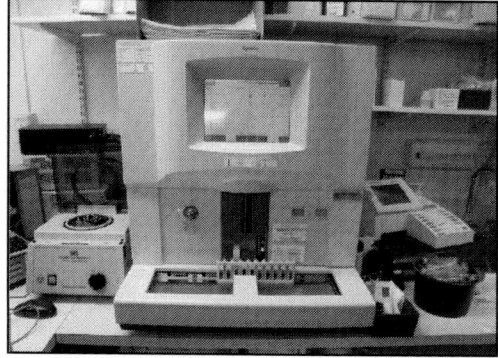

Figure 54   Hematology Analyzer.

- Urinalysis

As the name suggest, it is the process of examining the urine for the purpose of diagnosing conditions such as: urinary tract infections, kidney diseases, etc.

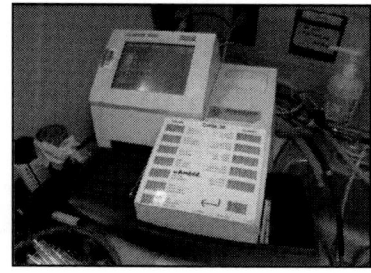

If you work with laboratory equipment, you will require special training and precautions because you can be exposed to fluids, tissues, pathogens and hazardous

Figure 55   Urinalysis Machine.

materials ("Biomedical Hazards"). Everyone is required to wear personal protective equipment (known as "PPE") to work with any laboratory device.

Generally, service repairs and preventative maintenance (PM's) to clinical laboratory equipment are provided by the manufacturers of the units. So, if you want to work with clinical laboratory equipment, try to learn more about the companies that manufacture this type of units. Also try to contact technicians that are already working in this field. (See the chapter titled "Networking" for suggestions).

4.11 Patient Monitors (Physiological Monitors)

Physiological Monitors can be found in almost all patient care areas as a "bedside monitor", whether in the surgery department, intensive care unit, or recovery rooms. Other models are portable, battery operated and are used when transporting patients. Others are designed to be used in neonatal care.

These units are used to view the patient's vital functions when constant monitoring is needed. A series of different sensors are connected to the patient (either in the chest, arms, fingers, etc.). The device is designed to amplify the signals, process them and display them on a screen (either as a wave or a numerical value or both).

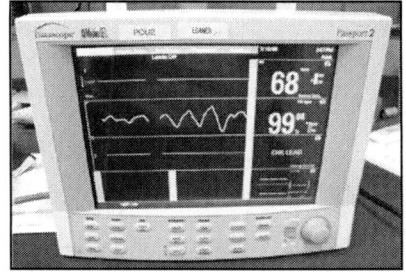

Figure 56 Modern Physiological Monitor

Depending on the brand and model of the equipment, a Patient Monitor can include features to measure physiological parameters such as:

- Temperature
- EKG (see Section 4.5)
- NIBP (Blood Pressure / See Section 4.2)
- Pulse (Heart Rate)
- SpO2 (oxygenation of the blood / See Section 4.12)
- Respiratory Gases
- Many other features!

An important feature that the units come with is a series of alarms that can be set to alert the medical staff about any problems with the patient's health. These alarms are activated if a parameter exceeds what is specified.

For example, an average heart rate of an adult can vary from 50 to 90 BPM (beats per minute). We can create an alarm condition that is activated if the patient's pulse is over 100 BPM, or if it goes below 50 BPM. That way, the device can tell us when a patient's pulse rises or falls to dangerous levels.

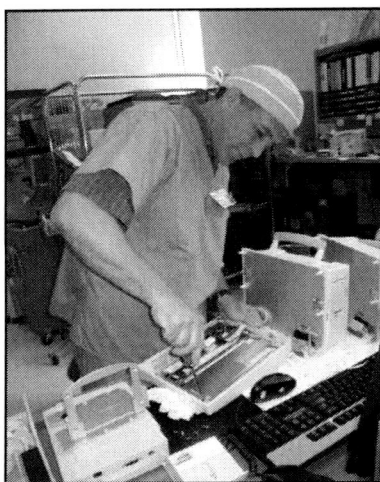

Figure 57   Biomed repairing a Datascope Patient Monitor.

Since these devices are used continuously, it is important to examine the condition of the batteries, sensors and transducers and the physical condition of the unit.

4.12 SpO2 (Blood Oxygenation monitors)

SpO2 machines, known as "Pulse oximeters" are used to measure the amount of oxygen in the blood of the patient. The unit gives the measured total as a percentage (%) and it also indicates the heartbeats per minute (heart rate).

Figure 58 Finger Sensor

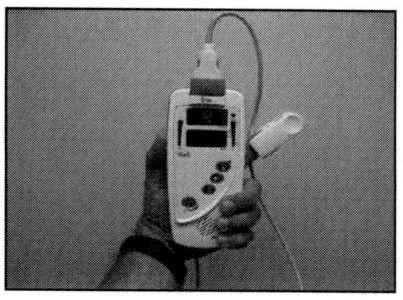

Figure 59 Portable SPO2 Monitor.

Normally, a sensor is placed on the patient's finger or on the earlobe. Each sensor has an LED (light emitting diode) and a photo detector on the other side. Part of the infrared light emitted by the LED goes through the skin (finger, ear, etc.) and the rest is absorbed by the tissue.

Each time the heart beats, the amount of light absorbed by the tissue changes. The light is absorbed in a different way when the blood is oxygenated (oxyhemoglobin) than when it is not.

The device finds these differences in absorption and calculates the percentage of hemoglobin that is oxygenated. That percentage is the number you see on the screen of the unit.

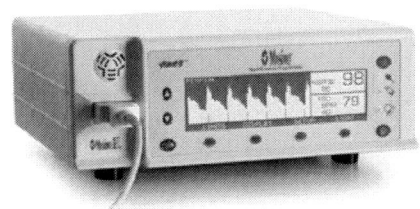

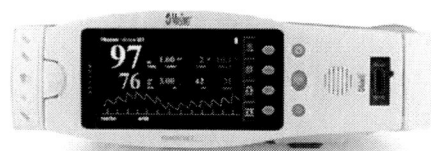

Figure 60 Stand Alone SPO2 units. (Courtesy of Masimo).

This machine can be found as an individual unit (portable or desktop) or as component of other equipment, such as NIBP monitors and physiological monitors.

Figure 61 shows some of these units.

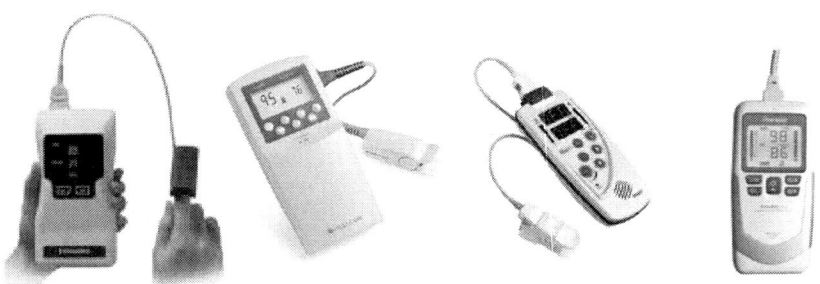

Figure 61   Portable SPO2 Monitors. (Courtesy of Nellcor, Masimo y Beful LTD).

4.13 Telemetry

Telemetry is a technology that transfers and receives data from a distance. Telemetry is a word derived from the Greek "tele" (distance) and "metron" (measure). **[18]**

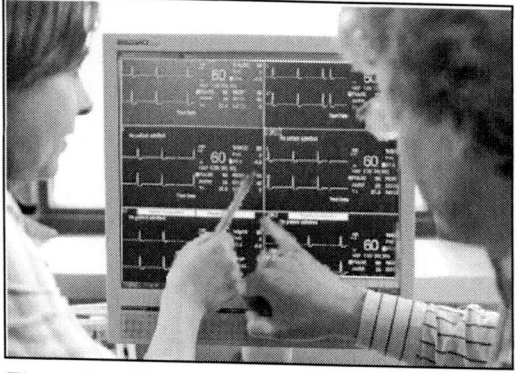

Figure 62   Telemetry System (Original picture courtesy of Philips).

This system works much like the patient monitors. Different sensors are connected to a patient to view the patient's vital signs on a monitor. The difference in a telemetry system is that the information is monitored in a central unit outside of the patient's room. What is the purpose? I will give you an example.

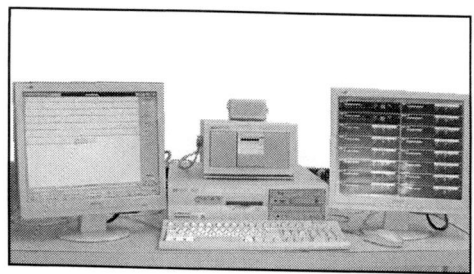

Figure 63   Central Telemetry Station.

Each patient that is in an intensive care room is connected simultaneously to different machines, like infusion pumps, ventilators, SpO2 machines and / or multiple vital signs monitors. Now imagine how difficult it would be for a nurse or a doctor on duty to visit each patient continuously in order to monitor their vital signs in that room. It would be an endless and exhausting task!

In a modern telemetry system, you connect the electrodes to the patient and they are attached to a transmitter known as a"Tele-Box". The unit is usually the size of a TV remote control, as you can see in Figure 64. This transmitter sends the acquired signals (vital signs, EKG, SpO2, blood pressure, etc.) to a central unit, with several displays.

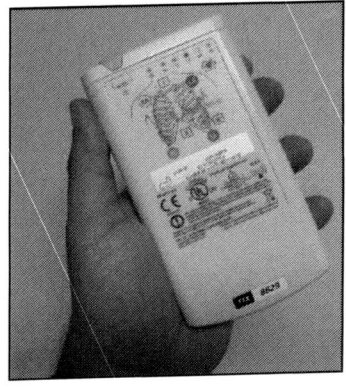

Figure 64 "Tele-Box". Note the size of the unit.

These central units are located in another room, where Telemetry Technicians (Tele-Techs) work. These technicians are dedicated to monitor all screens constantly. In an emergency, the system gives an audible and visual alarm that tells the Tele-Techs if a patient is having a problem. Immediately, the tech is responsible for notifying other personnel (doctors and nurses) to meet the emergency.

Figure 65 Telemetry Room. Tele-Techs constantly monitor the patients' vital signs.

Typically, up to 16 patients at a time can be monitored on each screen of a telemetry system, depending on the manufacturer specifications and the software licenses that are acquired by the department. Other systems provide even more patients per screen.

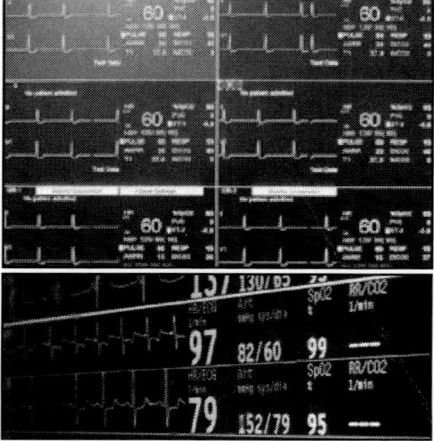

At the top of Figure 66 you can see how a Tele screen looks with 12 patients and the center pictures shows one with 6 patients. Each box represents a patient and you can see the patient's EKG signals, pulse and other vital signs from each patient.

Each system has printers to print the information needed for medical evaluations and an electronic filing system where all information is stored for later use.

The following pictures are various types of transmitters or "Tele-Boxes":

Figure 66   Various Telemetry screenshots.

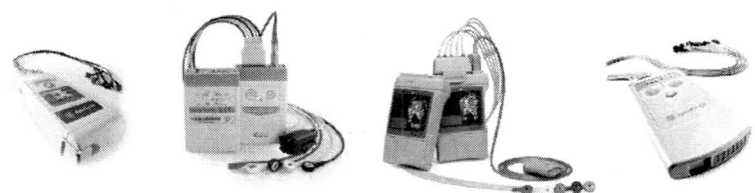

Figure 67   Various Tele Boxes. Some units are from GE/Marquette, Philips/HP and others.

4.14 Ultrasound

All the sound waves that are above the audible range of human hearing are called "ultrasound". It has been shown that humans can hear from 20Hz to 20KHz. So, all the frequencies above 20KHz are considered "ultrasound." In fact, "ultra" can be defined as "High", "Radical" or "Extreme."

An ultrasound wave is very similar to the echo created by bats which they use to navigate. The bat emits an ultrasonic sound, which travels up to the objects and "bounces" back. Then the bat captures the returning signal and manages to avoid obstacles. It's the same principle used in radar or sonar.

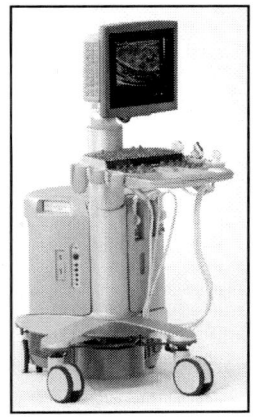

Figure 68a
Modern Ultrasound Machine. (Siemens)

The ultrasound machines are able to create an image with the "echo" of the returning waves from the tested areas. The machines usually transmit at a high frequency (from 1 to 10 MHz). The machine works with a probe that is placed on the skin surface and it sends the signal to the inside of the body. A conductive gel is used between the probe and the patient's skin to improve the contact and the signal transmission. The probe is moved over the surface of the area to be examined.

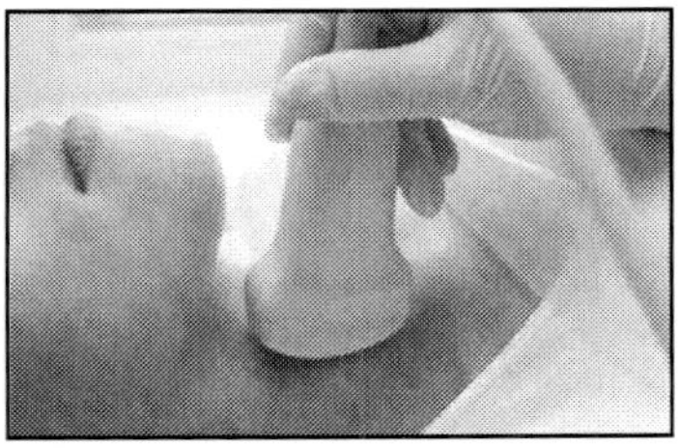

When the signal hits different types of tissue, (skin, bone or fluids), the signal "bounces" back to the probe, which captures it using different sensors. A computer calculates the time the signal takes from the probe to the tissue or organ and the time it takes to return.

Figure 68b Ultrasound Probe in use. (iStockPhoto.com).

From this data, the device creates a two-dimensional image that we see on the screen. That way we can see internal organs such as kidneys, the liver, the heart, the veins and arteries, and even a fetus.

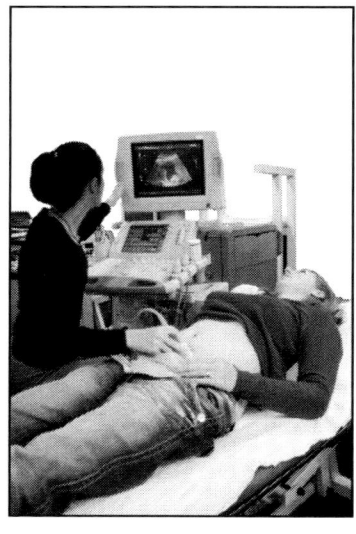
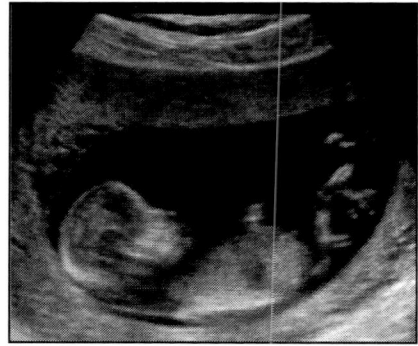

Figure 69　Ultrasound system and an image of a fetus as seen in the display. (iStockPhoto.com).

Ultrasound has many uses in medicine. Not only it is used to create images, but it can be used for physiotherapy. By sending a controlled signal to the muscles, you can heat the internal tissues in the body.

4.15 Ventilators

Ventilators are automated machines designed to help patients breathe. It "inhales and exhales" mechanically for the patients when they can not do it by themselves.

We know that humans breathe oxygen (O2) and exhale carbon dioxide (CO2). Therefore, the machine must help the patient do exactly the same: provide oxygen and remove carbon dioxide.

Figure 70 Puritan-Bennett 7200 Mechanical Ventilador.

Using pneumatic systems, electronics and compressors, the ventilator mixes oxygen (from tanks or from oxygen supply lines from the hospital system) with air at specified quantities and then administers the mix to the patient by a pipeline known as a "breathing circuit". This mixture is moistened and heated before reaching the patient. The machine pressurizes the mixture until the patient's lungs are filled, "forcing" inspiration. Then it releases the pressure for exhalation.

The machine has several controls to adjust the type of ventilation (manual, automatic or intermittent), the volume needed to inhale (tidal volume), the concentration of oxygen, air pressure, etc.

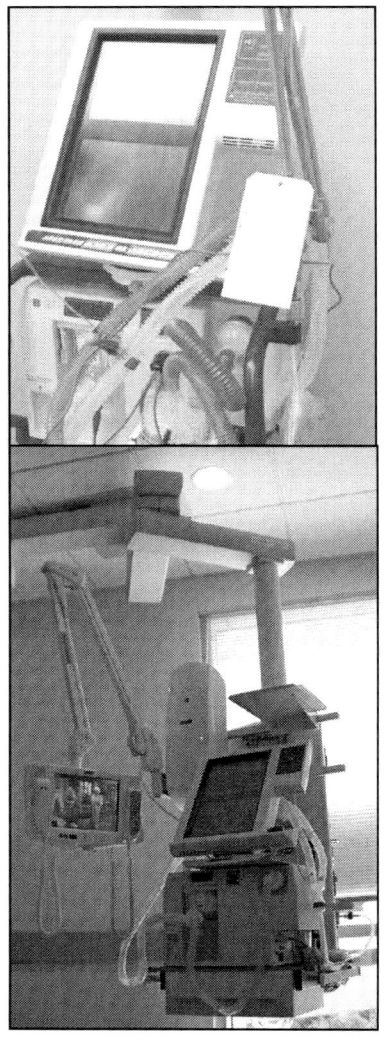

Like other biomedical equipment in this book, the ventilators are considered *"Life Support Equipment"*. All of them come equipped with a series of adjustable alarms which will sound immediately if a medical problem is noticed with the machine or the patient.

You should not work on these devices without the proper training. The calibration and adjustment of this type of equipment is crucial to patient's lives.

Figure 71 Modern Puritan-Bennet 840 Ventilator, and the same unit installed in a ceiling mount in an ICU room.

4.16 Radiographic Equipment and Digital Imaging

X-ray systems are an invaluable tool in the medical field because they are used to see inside the human body without resorting to surgery or other invasive procedures.

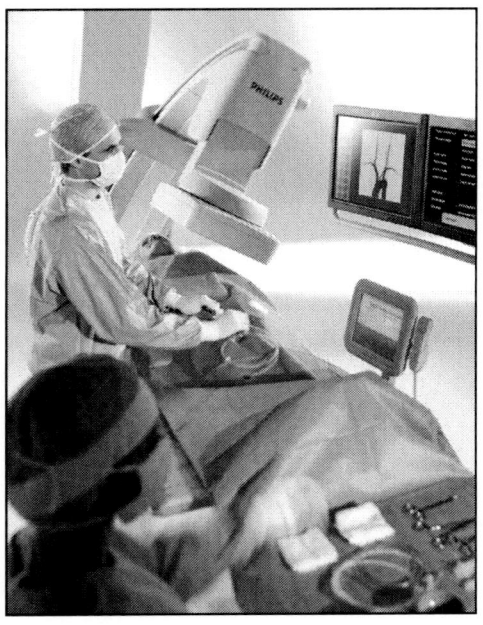

Figure 72 Surgical procedure using Radiographic equipment. (Philips).

Since its discovery, the X-ray and X-ray based equipment are considered *"the discovery of more rapid distribution of all times"* **[22]**.

Over the years, this technology has evolved and led to the creation of new computer systems that have completely revolutionized the field of medicine.

There are entire books devoted to this broad field of medicine, so I will only give you a basic idea of how these devices work. The operation and theory will be explained in detail by your Biomed teacher.

4.16.1 X-Ray Machines

As many of you already know, an X-ray machine is used to take "pictures" inside the human body, called "radiographies" or simply "X-rays." There are several types of procedures that use X-rays, such as angiography, fluoroscopy, mammography and CT Scanning.

The "X-Ray" was discovered in 1895 by a German physicist called Wilhelm Roentgen [22]. He named his discovery as "X", because those rays were unknown to him! Roentgen noticed that this type of radiation had the ability to pass through many materials while being absorbed by others.

Figure 73   Doctor examining a radiography image (Courtesy of Smith & Nephew).

Figure 74a X-Ray Tube. (Courtesy of iStockPhoto.com).

The centerpiece of this machine, known as an "X-ray tube" is vacuum-sealed and has two electrodes, known as the anode and cathode (similar to a diode). This X-ray tube is placed on one side of the patient and a photosensitive film is placed on the opposite side.

Within the X-ray tube, a device "fires" high energy electrons (from the cathode) into a rotary target (anode). These electrons hit the target and produce photons (light), but in a much higher energy range than visible light.

This energy is what we call "X-Rays." These rays are directed toward the patient.

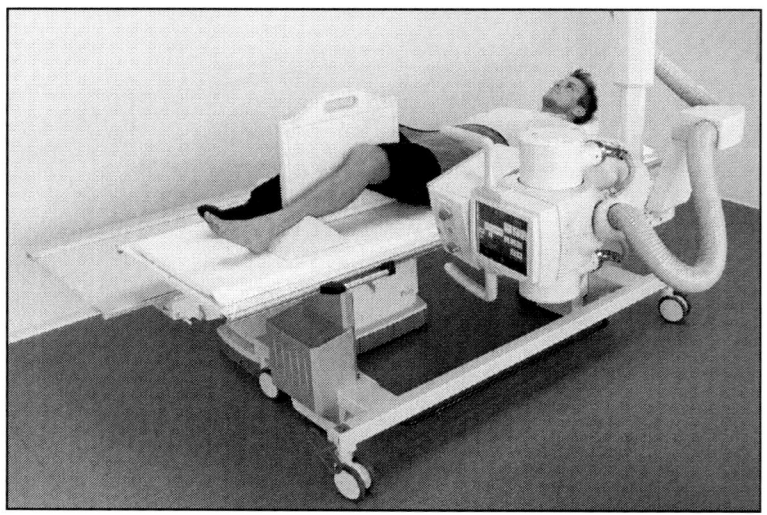

Figure 74b   Patient in X-Ray Room.   (Courtesy of Siemens.)

What's interesting about this process is that the human body has different types of tissues of different densities. The skin, muscles and internal organs are soft tissue, while bones and tendons are much more dense material. X-rays pass through the skin and soft tissue, but are absorbed by the hard tissues and bones. This leaves an impression in the film.

Similar to developing a photo, the radiologic film is less exposed in the areas where x-rays are absorbed by the hard tissue and they look clearer. This gives an excellent picture of the bones and tendons in this part of the body.

It is important to use this technology with caution. X-rays can be dangerous because they are a form of "ionizing radiation". It can cause burns, chemical reactions in cells, and even affect the DNA causing mutations, cancer or damage to a fetus.

The technicians who repair these units must be very careful about how much they are exposed to these rays while providing service. The technician could be exposed to lethal amounts of radiation if safety measures are not taken.

Moreover, this type of equipment tends to be large and heavy. Normally a repair technician must seek help to move this equipment or while disassembling it. If you are interested in working with X-ray equipment, remember that security is of paramount importance, so never work with an X-ray unit without having the necessary training.

4.16.2 CT Scanner (Computerized Tomography)

A CT scanner works much like an X-ray machine. The main difference is that the rays are projected from different angles, (the tube rotates around the patient) and it does not use film. The whole process is done by a computer system.

X-rays emerge from one side of the system, while receptors or sensors on the other side send the captured images to a computer system.

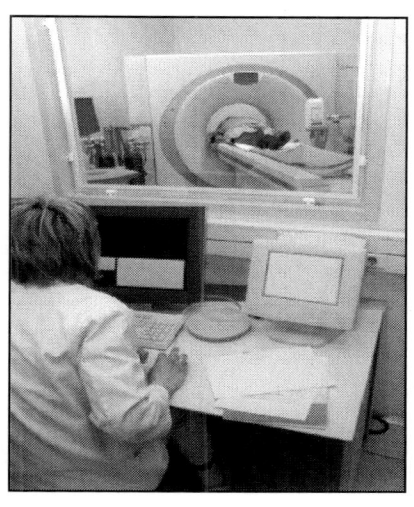

Figure 75   CT System. (iStockPhoto.com).

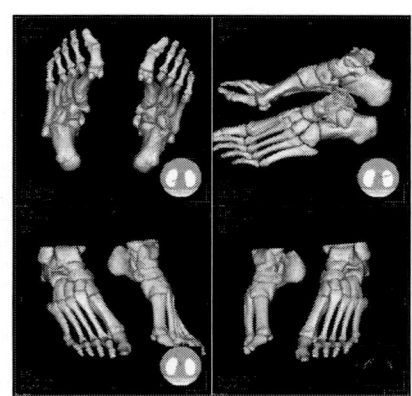

Figure 76   Computerized Tomography shows your internal organs, bones and tissue in 3 dimensions (3D). (Courtesy of iStockPhoto.com).

The system can take hundreds of different shots from different angles. The computer system is used to assemble the views and creates a three dimensional image (3D).

This technology allows you to see full organs or a specific body section. Since the computer is able to generate 3-D images, they can be manipulated (rotated, sectioned, etc.) by the computer system.

This technology is extremely useful to diagnose tumors and other conditions.

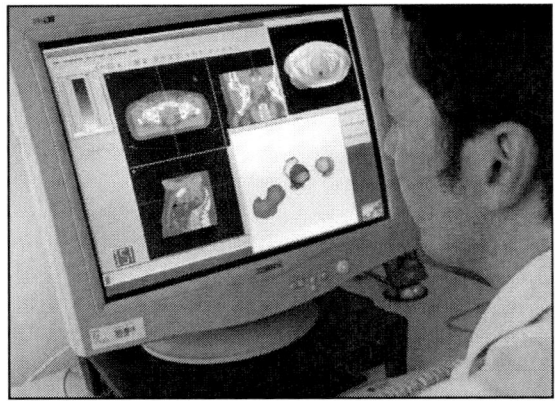

Figure 77a Computerized Tomography process. (Courtesy of Philips).

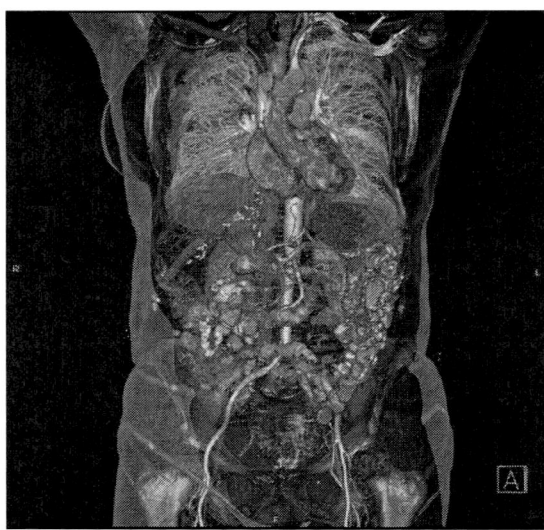

Figure 77b 3D Tomography image. (Courtesy of Siemens).

4.16.3 MRI Machines

The Magnetic Resonance Imaging (MRI) equipment is very similar to the Computerized Tomography (CT Scanner), but there is a big difference ... it doesn't use X-rays!

The MRI uses very strong magnets, *"7,000 times stronger than the magnetic force of the earth"* **[29]**.

The MRI machine has a circular shape and the patient enters through a central opening. But how is the test done?

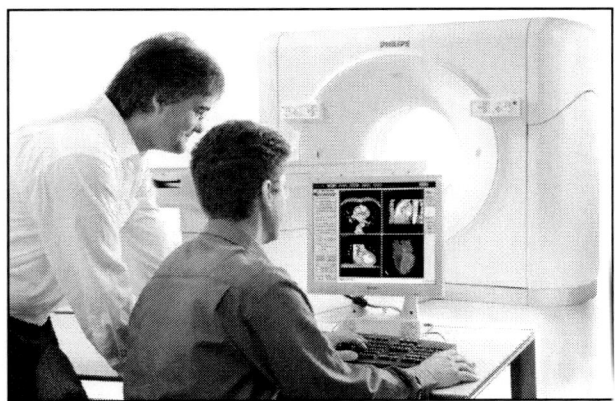

Figure 78   MRI Machine. (Courtesy of Philips).

The hydrogen atoms in the body of all patients react to the magnetic fields created by the MRI machine **[29]**.

One of the easiest explanations I've found about how an MRI machine works is by an engineer named Rodolfo Gutierrez.

He explains the process in his personal online blog (translated from Spanish):
(http://tecnologiadeimagenesdiagnosticas.blogspot.com):

"The human body is made up of billions of different atoms, which are the basic blocks of everything. Hydrogen atoms are the ones that are needed for the MRI test because they have a magnetic presence that is captured only during the scan. The amount of hydrogen atoms in the human body is what creates the anatomic images. This is the resonance part. The MRI machine applies radio frequency pulses to the points you want to analyze and this causes the atoms to absorb energy and turn or spin.

When the radio frequency pulse excitation is turned off, the protons begin to rotate more slowly, returning to normal, and releasing magnetic energy. The amount of energy liberated by these atoms reaches the computer system. This information is digital and becomes an image that is presented on a screen." **[30]**

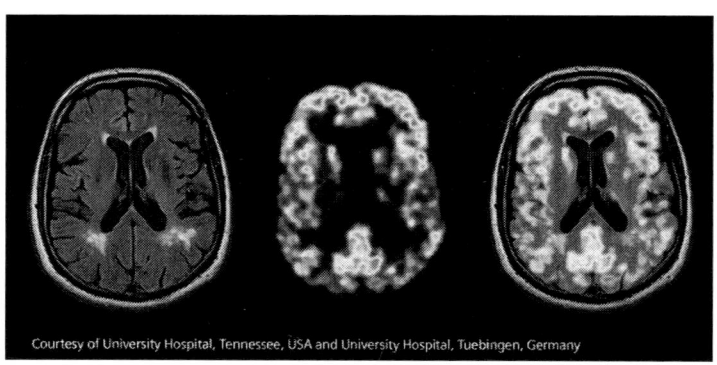

Figure 79   Magnetic Resonance Images (Courtesy of Siemens).

This computerized system provides highly detailed images of the inside of the body without causing hazards like the X-rays do. In addition, the computer allows the user to rotate sections, split up and color the images, to have a clearer picture in order to make a correct diagnosis.

The MRI is used to identify tumors and other formations in the head, spine, chest, abdomen, pelvis and extremities. Often, this type of diagnostic information can not be acquired with any other medical procedure, except surgery. **[29]**

So, the MRI has been one of the inventions that has changed the course of modern medicine!

Part 5:
Tools, Equipment Repair and the Biomed Department

If you're already studying Biomed or Basic Electronics, I infer that you know basic electricity and electronics concepts (Ohm's Law, series and parallel circuits, capacitance, semiconductors, etc.). Plus, you have a great disposition to learn new stuff. Excellent!

The first thing I will tell you is that the theory you learn as a student is somewhat different than the practice.

In electronics theory, we always speak of electronic components that are "perfect" or "ideal" and in practice, that *does not exist*. All electronic components have a life expectancy, are prone to be affected by noise and external signals, by beating (abuse) and the technical specifications can vary with time, temperature or voltage variations.

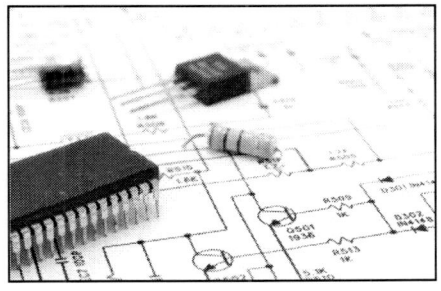

Figure 80 You will notice that there are differences between Theory and Practice. (Courtesy of iStockPhoto.com).

Something you will probably notice is that a Biomed technician usually does not work at component-level repairs (replacing resistors, diodes, capacitors and so on). We usually replace the board (or card) that is defective. This is because the regulatory agencies (such as FDA) prevent component-level repairs. Why?

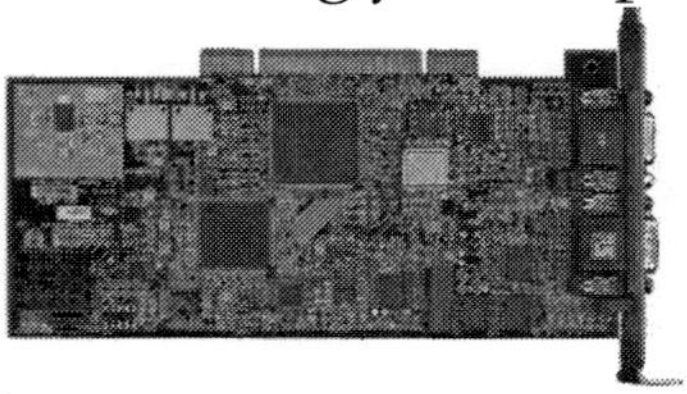

Figure 81 It's necessary to replace defective boards only with original parts.

99

Imagine what could happen if we replace a resistor or other component with a non-original part or of inferior quality! If we do that, we would be modifying the device and creating unnecessary risks to patients' lives. That is why the regulatory agencies (such as FDA) specify that we use only original cards, parts or modules produced by the original manufacturer and certified by the agency.

Now the question is, what are the tools and devices needed to repair Biomed equipment? Check this in the next section.

5.1 Basic Tools

To repair biomedical equipment the basic tools are very similar to those used for repairing electronic equipment. These include among others, the multimeter, screwdrivers of all kinds (including "torx") and pliers, soldering iron or welding station, and so on. Sometimes, an oscilloscope is necessary for critical adjustments.

Figure 82   Common tools used in Electronics repairs (Courtesy of Velleman Products).

Depending on the type of service that the unit requires, it is desirable to have the hand tools and test instruments required for each step of the repair process. We want to make the repair the best possible way.

Figure 83   Portable Oscilloscope. (Courtesy of Velleman Products).

Are there differences in the equipment used to repair medical equipment vs. other electronic devices?

Yes. In the case of biomedical equipment there is a huge diversity of test equipment, such as "Physiological Simulators" Now you may wonder, What is a *Physiological Simulator*?

A Physiological Simulator is a piece of test equipment used to replace a human being when you are testing biomedical devices. Obviously, it would be unwise to test certain equipment on a human being. Imagine what would happen if you tested a defibrillator, firing the electrical charge on the chest of another person! Or what would happen if you tried an anesthesia machine on a coworker? We could cause irreparable damage or even death to another person!

In this chapter I will show some of these devices and how they are used.

5.2 Physiological simulators and test equipment

As mentioned in the previous section, there is a large variety of tools for testing and repairing medical equipment. All equipment that is used to examine other medical devices must be calibrated correctly *, as *the proper functioning of medical devices and the lives of other human beings depend on this.*

In this section I will go over some well-known test equipment in the Biomed field. Obviously, this section does not cover all the test equipment that exists because each medical device and medical specialty have their own testing equipment.

The following are just a sample of some equipment that I use on a daily basis to test, calibrate and repair medical equipment:

- "Safety Analyzer"
- EKG Simulator
- Electrosurgery Tester (ESU)
- Defibrillators Tester
- NIBP Simulator

Now let's see how these devices are used.

(* Note: Governmental agencies are responsible for creating the "Standards" with which we compare our test equipment, to know how accurate they are, or how well "adjusted" or calibrated they are. Check the section of Regulatory Agencies / Part 6 for more information.)

5.2.1 "Safety Analyzer"

One of the main tools used by Biomed Technicians is the "Safety Analyzer", or Electrical Safety Analyzer. This equipment is used on a daily basis by all technicians. I will elaborate on this device:

A Safety Analyzer is used to verify the electrical status or electrical condition of the medical equipment we are checking. Can this test be performed using a normal multimeter?

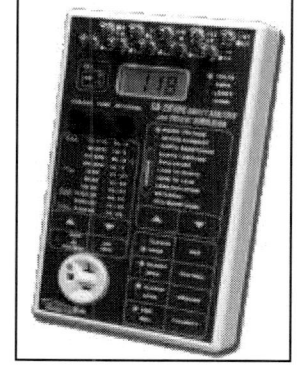

We know that the multimeter can measure resistance, current and voltage. However, the Safety Analyzer does much more. Remember in the Electrical Safety section we talked about "Leakage Current"?

Figure 84   Safety Analyzer (Courtesy of BC Biomedical).

Figure 85 MedTester 5000C. (Courtesy of Fluke Biomedical).

The Safety Analyzer is specially designed to find this type of problem. For example, the analyzer can tell us if there is an internal misconfiguration of the power cord cables of the unit under test! In fact, the chassis-to-ground resistance of any medical equipment should be 0.5Ω or less **[11]** to avoid problems of leakage current. The Electrical Safety Analyzer tells you if the device meets these test specifications. Other manufacturers add other options to their Safety Analyzer, as an integrated EKG tester, or internal memory chips for recording the data, and other useful features.

What other kind of Biomed test equipment exist? Read on.

5.2.2 EKG Simulator

Figure 86   EKG Simulator. (Courtesy of Fluke Biomedical).

When you need to perform preventive maintenance on an Electrocardiogram (EKG) machine, instead of connecting the electrodes to the chest and limbs of a patient, we have an *EKG simulator*. This equipment simulates the operation of a human heart (heartbeat).

The unit has connectors to simulate each of the sensors that are placed on the chest of a patient. The device can create a normal heart signal or you can vary the number of beats per minute (BPM) to create 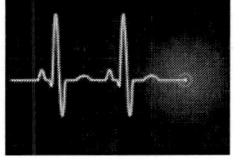 "tachycardia" and/or a "failure" to determine whether the equipment under test is functioning properly.

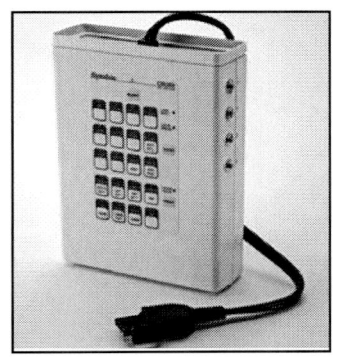

Figure 87   EKG Simulator. (Courtesy of Physio-Control).

It is a very easy to use unit because each connector is color coded for the corresponding wires on the EKG machine. It is also portable, so you can take it with you to any department where an EKG machine is used.

5.2.3 Electrosurgery (ESU) Tester

The Electrosurgery Tester (*ESU Tester*) is a very useful unit. As we learned, the ESU is used to cut and coagulate tissue. It is illogical to use a human being to check an ESU unit. Moreover, it would not be convenient to get a piece of meat to test the equipment either! How can we test an ESU to see if it is working properly?

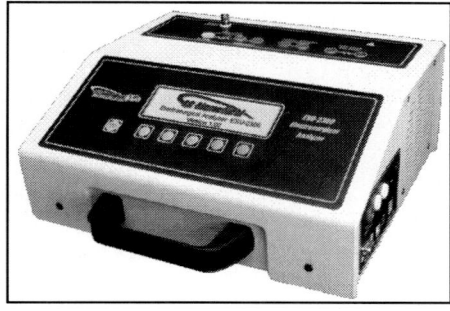

The Electrosurgery tester has been designed for this. This equipment measures the power (in Watts) that the ESU machine provides with internal resistive loads and electronic circuitry.

Figure 88   ESU Tester. (Courtesy of BC Biomedical).

For example, if you select *"40 Watts"* in the ESU control, the ESU tester should measure about the same amount of power. If the Tester's display shows a different result, it suggests that the ESU Unit is out of calibration or it has another failure.

The ESU Tester is easy to use. It is important to use the recommended resistive load to test the unit. (This information is supplied by the manufacturer).

After repairing an ESU unit, we use the tester again to confirm that the unit is in excellent working condition.

In Figure 89 you can see how I used an Electrosurgery tester to adjust the output power of a Phaco Machine * which is used for eye surgeries.

Figure 89   ESU Tester used to test a Phaco Emulsifier Machine.

It is important to calibrate the equipment following the recommendations of the manufacturer and the regulatory agencies. Remember that people's health or lives are at stake!

* (Phacoemulsification or Phaco - is a medical term for "cataract surgery with small incision." During the procedure, a small incision is done on one side of the cornea of the patient and the surgeon inserts a small probe. The probe emits ultrasonic waves to soften and break up the cataract, which is "sucked" out of the eye. Then they insert an artificial intraocular permanent lens). [31]

5.2.4 Defibrillators Tester

Other piece of essential test equipment is the Defibrillator Tester (or Defib Analyzer). This unit usually has 2 metal surfaces onto which the defibrillator paddles are placed in order to "fire" the electrical load safely and to measure the power output.

The unit has internal resistors and circuits designed to withstand the electric shock, while microprocessors analyze the data and display the results. It displays the delivered energy in Joules. Some models even have the ability to analyze pacemakers and come with an integrated EKG analyzer in the same unit. Also, some has special connectors to transfer data to computers or other equipment.

Figure 90   Various Defib Testers. (Courtesy of BC Biomedical & Fluke Biomedical).

As you can imagine, the annual calibration of these devices is critical, because defibrillators are used throughout all patient care areas.

5.2.5 NIBP Simulator

To test electronic sphygmomanometers, we use a NIBP* Simulator. It must "behave" as a human being under test!

Basically, these simulators operate as follows:

The cuff that is normally placed in the patient's arm is placed around a molded plastic part and the rest of the pipeline is connected to the sphygmomanometer *through* the simulator. Then, the testing begins…

While the NIBP Machine is doing the process of measuring blood pressure (filling the pressure cuff with air and releasing the pressure slowly), the NIBP Simulator imitates the pulsations of the patient's blood while monitoring the NIBP "behavior". That way, the simulator makes sure that the NIBP machine is measuring the pressure accurately.

In addition, the NIBP tester indicates whether there is an air leakage, which helps us to verify the condition of the air tubing and the cuff.

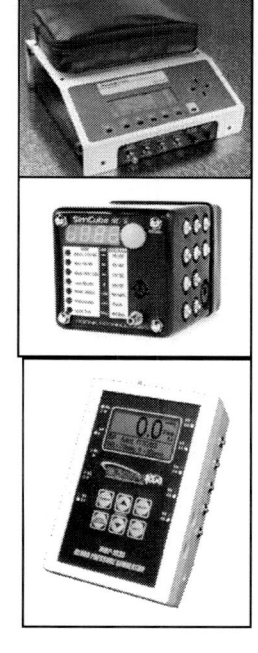

Figure 91   NIBP Testers. (Courtesy of Fluke Biomedical, Pronk Technologies & BC Biomedical).

* Always remember that NIBP means "Non-Invasive Blood Pressure".

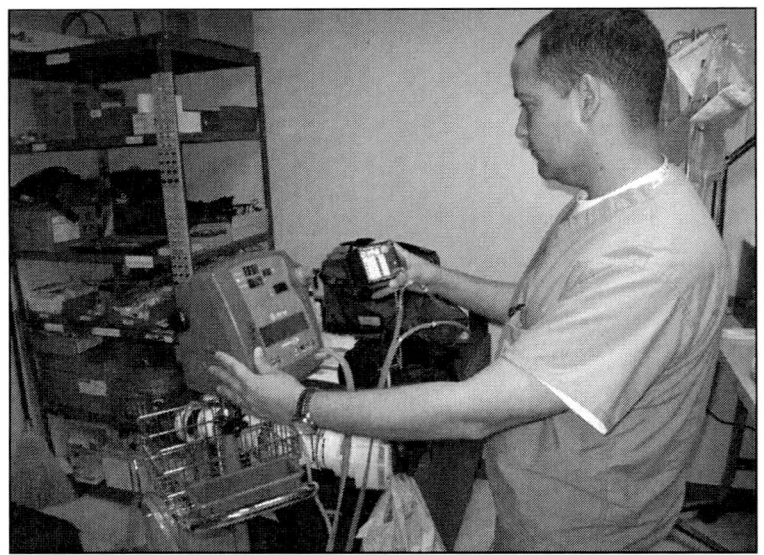

Figure 92a The Author testing an electronic NIBP machine using a "SimCube" tester from Pronk Technologies.

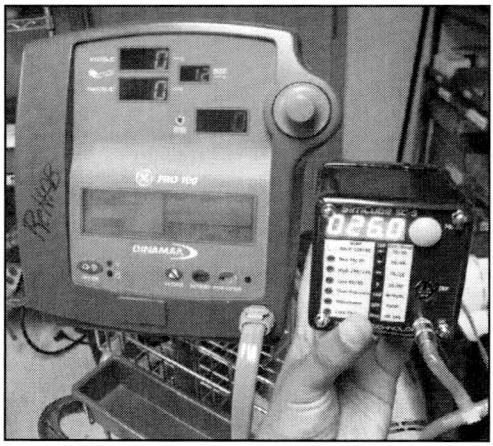

Figure 92b Front View of the units.

Although we have talked about a lot of medical equipment and test equipment, where do we perform the repairs? How do we do it? And how do we keep track of the repairs? The answers to these questions are in the next section.

5.3 The Clinical Engineering Department (Biomed Department)

The Clinical Engineering Department or Biomed Department in a hospital environment is in charge of providing proper maintenance to all biomedical equipment at the health center. This department is not only responsible for the equipment's maintenance, but the Biomed staff also works with the hospital administration to evaluate devices that the facility wants to acquire and coordinates all matters related to the installation of medical equipment in the hospital. It is much more than just equipment repairs! It is an interesting and dynamic field!

To achieve all these goals, the Biomed department usually has enough space (a shop or office), where workbenches are placed for the technician's repairs. In addition, there are tools and equipment needed for the repair and maintenance of medical devices, service manuals, repair parts in stock, etc.

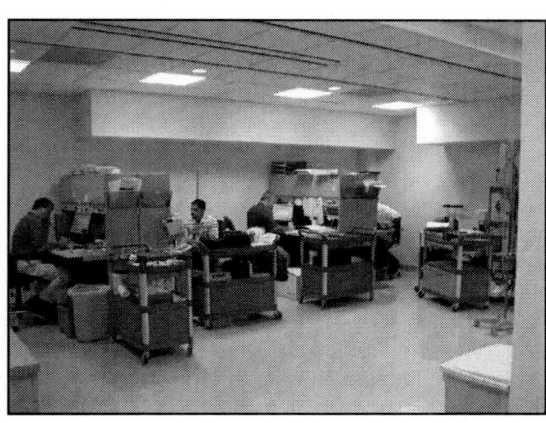

Figure 93   Modern Biomed Shop. Baycare Health System.

Additionally, the Department has files with the documentation of all existing equipment in the hospital!

These files and documents exist because it is required by law. Biomeds must keep accurate details of each biomedical unit in the hospital: brand, model, serial number, details about the warranty and records of each one of the repairs or preventive maintenance performed on each unit. You are probably wondering why you need to keep so much information?

Let me give an example. There may be a case where there is an incident with a unit involving a doctor and a patient or user. Perhaps the device failed and a patient is affected. Obviously, in an incident like this, responsibilities will be assigned in order to investigate what or who caused the accident (failure of equipment, programming error, user error, etc.).

There must be documentation showing that a technician has performed preventive maintenance (PM) on the unit within a specified timeframe. That way, if necessary, you can send information to the manufacturer about a failure you have found, for correcting the problem. Also, by having detailed facts about every PM or repair, the Biomed department can legally defend itself against any claim that the incident was due to lack of maintenance.

I will give you another example. Imagine that an incident occurs in which an infusion pump administers a dose of medicine higher than required for the patient ("over-infusing"). If the preventive maintenance has been done to the device on time and there are accurate records on file, you can prove that the equipment was properly calibrated and was working fine until the time of the incident. That way, we could rule out the lack of maintenance as the cause of the problem and identify that perhaps the failure was due to human error (user), or an unexpected failure of equipment.

Hospitals use modern technology to keep files or records updated. Many Biomed Shops have computer programs and systems in place for their inventory and for archiving the repair and maintenance records. Since there are hundreds or thousands of devices in a hospital, many inventory systems are being developed, such as using bar code stickers:

Figure 94 Barcodes. (Courtesy of GE Healthcare and Baycare Health System).

How do we use bar codes in modern hospitals? A number is assigned to each and every medical device in the hospital and tagged with a barcode sticker. All this information is stored in the computer system of the Biomed department. Then each Biomed technician is assigned a PDA ("Personal Digital Assistant"). Usually it is a small portable computer that has a built-in scanner.

Figure 95 PDA's.

What is the function of the PDA? The PDA has a copy of the entire inventory of the biomedical equipment at the hospital in its internal memory. That way, if you find any device anywhere in the

Figure 96 Identifying a unit by scanning the Barcode with the PDA.

hospital, you can "scan" the barcode on the unit and the entire history of the device will appear immediately on the screen of the PDA!

That means that you can immediately know details such as:

- Purchase date of the unit. Is the unit still under the manufacturer's warranty?
- When was the last repair or PM performed?
- What parts were replaced?
- Information on the due date of the next preventive maintenance service.

At the end of each working day, every technician connects their PDA to the server and all the information is updated. Having all the information loaded to the server allows us to see the full record of any device at any moment. We can see it on the computer screen or we can print a report with all the necessary data. This is very useful, especially when the hospital is surveyed by inspectors from a regulatory agency.

I'll give you more interesting details about the Biomed Department in a hospital:

It may be developed 2 different ways: "in-house" (internal) or "outsourced" (foreign company). What are the differences between each of these ways?

An "In-House" Biomed department is organized by the hospital administration. The Biomed engineers and technicians are *hospital employees*. Among other things, the technicians have the advantage of training/repair classes directly from the equipment manufacturers. On the other hand, sometimes it is difficult to get information and service manuals from some manufacturers and the hospital has to bear the responsibility of the ordering and shipment of parts, equipment, etc.

In addition, the "In-House" department is responsible for the compliance with all requirements from the hospital administration, including budget limitations and so on.

In the case of an "Outsourced" Biomed department, the hospital hires an outside company * to do everything that a normal Biomed Department is assigned to do. Transferring these responsibilities from the hospital to a hired company proves beneficial to some institutions. If you are planning to relocate, working for an "outsourced" biomed company allows you the opportunity to continue working for the same company at a different city.

On the other hand, the contracts between a hospital and any external company have a time limit. So if you work for an "outsourced" company, your job at the hospital depends on whether the hospital will continue to renew the contract with the company you work for.

The outsourced company has to comply with all the terms of the contract, but must comply also with government laws and regulations designed to ensure quality of service and safety requirements.

What are some of these agencies? Let's see in the next chapter.

* (Note: Some well-known companies that offer this type of service are GE Healthcare, Aramark, and Philips).

Part 6:
Regulatory Agencies, Accreditation and Legal Aspects

As I have explained in the past section, hospitals are "surveyed" or examined regularly by various independent agencies or the government. These regulatory agencies are responsible for "monitoring" the institution, to see if they are complying with all the legal requirements and to help maintain the highest quality of services offered by the hospital. Therefore, the Clinical Engineering Department (or Biomed) must also comply with certain regulations.

Now I will mention some of these regulatory agencies or "Boards" with which you will be involved if you work in the United States or Puerto Rico. Some of them perform unannounced inspections to the hospital facilities. I'm also including in the list information about some governmental programs and laws that you will hear about that are involved in the healthcare field.

6.1 Joint Commission for Accreditation of Health Organizations (JCAHO):

- It is a nonprofit organization and is the most well known of the accrediting agencies in the U.S.

- All hospitals want to pass the JCAHO accreditation. Why? Because it is a requirement in order to collect reimbursements from Medicare and Medicaid.

- JCAHO is in charge of surveys in all hospitals, to assess whether the institution is complying with federal regulations.

For more information, visit the official website: http://www.jointcommission.org.

6.2 Medicare and Medicaid

- Are US government programs created in 1965 with the idea of providing affordable health insurance to people of 65 years or more and some with certain disabilities.

- One of the main sources of health care reimbursement in the United States.

- Official Website: http://www.medicare.gov

- More information: http://www.cms.hhs.gov

6.3 FDA - Food and Drug Administration

FDA U.S. Food and Drug Administration

- The Department of Food and Drug Administration (FDA) is responsible for protecting public health, ensuring the effectiveness of the drugs that go to the market, organic produce, among others. It also regulates all medical equipment that makes it to market. Their website states:

"The FDA is responsible for protecting the public health by assuring the safety, efficacy, and security of human and veterinary drugs, biological products, medical devices, our nation's food supply, cosmetics, and products that emit radiation. The FDA is also responsible for advancing the public health by helping to speed innovations that make medicines and foods more effective, safer, and more affordable; and helping the public get the accurate, science-based information they need to use medicines and foods to improve their health". [32]

- No medical device can go to the market without the FDA approval
- All medical devices must meet requirements and Manufacturing Standards imposed by the FDA
- The FDA maintains a database of reports of medical equipment that fail or cause adverse reactions. Thus forcing manufacturers remove failed equipment from the market ("product recalls") and correct mistakes before allowing the product back in the market.
- In addition, they constantly publish reports about equipment or manufacturers that have violated FDA regulations or equipment that present a health risk to users. They are known as "Recall Alerts". As a Biomedical Technician, you must be aware of these reports constantly.
- More information: http://www.fda.gov .

6.4 OSHA - Occupational Safety & Health Administration

- OSHA's mission is to prevent work-related injuries, illnesses, and deaths. OSHA's websites explains this: *"The mission of OSHA is to ensure the safety and health of workers in America by establishing and enforcing standards, providing training and education, establishing partnerships and encouraging continuous improvement in occupational safety and health in the workplace"*. [34]

- OSHA establishes protective standards, enforces those standards, and reaches out to employers and employees through technical assistance and consultation programs.
- They conduct inspections in thousands of workplaces annually.
- Official Website: http://www.osha.gov .

6.5 AHCA - American Health Care Association

- It is a nonprofit association founded in 1949, with links to various state health organizations in the United States.
- It represents more than 10,000 health care providers
- Provide care to more than 1.5 million seniors and individuals with other problems.
- Official Website: http://www.ahcancal.org.

6.6 NFPA - National Fire Protection Association

• This Association was created in 1896, and is the "authority" in Electrical Safety, Fire and Construction in the United States. NFPA develops, publishes, and disseminates codes and standards intended to minimize the possibility and effects of fire, electrical risks and others. [35]

• NFPA has several "standards" or rules regarding the use of medical equipment in hospital facilities. The rule NFPA99 is well known to biomed technicians. It specifies the criteria to be followed to minimize the risk of fire, explosion or electrical hazards in the hospital facilities.

• Official Website: http://www.nfpa.org .

6.7 UL - Underwriters Laboratories

• For over 100 years, this has been the agency responsible for testing all types of products to confirm that they are safe for use by the general public before entering the market.

• UL Standard 60601 protects the patients, because it covers all the general requirements for the use of electro-medical equipment, including:
- Electrical Safety
- Spills
- Use of flammable anesthetics
- Sterilization Methods

• Official Website: http://www.ul.com.

6.8 NIST - National Institute of Standards and Technology.

- Federal Agency Established in 1901 with the goal of "promoting innovation and industrial competitiveness through technology and standards." **[36]**
- NIST respond to industry needs for measurement methods, tools, data, and technology by providing those measurement methods, instrumentation, and creating all measurement "standards".
- How does it affect the Biomed field? : All test equipment we use should be calibrated by comparing them with a standard developed by NIST, or have "tracking" to NIST ("NIST traceable"). This means that there must be documentation that indicates that our test device was calibrated by a unit originally calibrated by NIST or a NIST Standard.
- Official Website: http://www.nist.gov .

6.9 HIPAA - Health Insurance Portability and Accountability Act

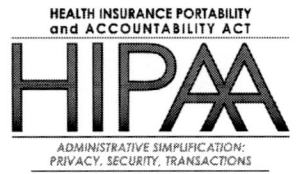

- It is a law enacted in 1996 by the US Congress. *"HIPAA protects health insurance coverage for workers and their families when they change or lose their jobs. It also requires the establishment of national standards for electronic health care transactions and national identifiers for providers, health insurance plans, and employers. It helps people keep their information private".* **[37]**

- HIPAA is well-known for the restrictions and regulations regarding the security and privacy of patient information
- More information: http://www.cms.hhs.gov/HIPAAGenInfo .

Perhaps at this point, it is a little confusing for you to hear about the many agencies and all these legal processes. However, all of this will become part of your daily work and you will be doing your job while taking into account the regulations and rules that are established by these agencies by default. Visit the official sites to get familiar with this information.

Part 7:
Your Career as a Biomedical Technician

As you may have noticed, the Biomed field is huge and there are many alternatives or many branches in which you could specialize. From Laboratory to General Biomed and/or Digital Imaging, there is <u>a lot</u> to learn!

No matter what area you decide to work in when you graduate, what can you do to develop and improve your skills in this field? Are there advancement opportunities?

In the next sections, I'll give you some suggestions on what you can do to keep "up to date" in your specialty and to help you to grow as a Biomed professional.

7.1 Professional Certifications

Some people think that a "certification" is only something similar to a "diploma" that you hang as decoration on a wall. But is it just that? What is a Certification? Is it necessary?

A certification shows that you have passed a series of tests, which validate your knowledge and skills in a particular area. To achieve a certification, you must pass difficult exams and it *is not an easy task*. It involves hours of study, effort and dedication.

Large and well-known companies like Microsoft and many other Associations offer tests that validate your knowledge about programs (software) or services that the companies represent.

In the healthcare field, many professionals are required to be certified. For example, in nursing, an "RN" Certification (Registered Nurse) is REQUIRED to carry out their work. In the case of Clinical Dietitians (Nutrition Specialists), the "RD" Certification ("Registered Dietitian") is required to perform their job.

In all technology fields, such as computers and networking, there are all kinds of certifications as well. Most are not "compulsory" in order to do your job, but getting one or more certifications could make a big difference in your career.

Microsoft offers dozens of certificates for their products and services. I will show you (using *Microsoft* as an example) how obtaining a certification can be a positive step for your career:

There is a Microsoft certification called MOS ("Microsoft Office Specialist" *). If you pass the exams and obtain this certification, you will be considered *an expert* using the Microsoft Office Suite (Word, Excel, Powerpoint, Outlook, Access and Project). You might ask: What benefit can I obtain if I acquire a certification like this?

If you get this certification, and you include this in your resume, it will be easier to assure a future employer that *you have the necessary computer and software knowledge to perform the job* required, even before the recruiter asks you!

In fact, an employer will know that you can start working immediately, and would not have to worry about providing software training for you, if he knows that you have the certification: He already knows that you are an expert in the use of these programs!

Other well-known certifications are the *"A +"* and *"Network +"* offered by ComTIA (http://certification.comptia.org) for computer and networks technicians.

Figure 97 CompTIA Logo.

* (Note: You can view details on this certification in the official Microsoft Web Site:
http://www.microsoft.com/learning/mcp/OfficeSpecialist)

CompTIA's website states that *"Certified professionals report that they have more confidence on the job, see an improvement in the quality of their work and have a higher demand for their services. Employers feel the same way. In fact, 74 percent of IT managers say CompTIA certifications are an important factor in considering an employee for a promotion"*. **[19]**

Taking all this into consideration, should you become certified in your field? Why? How? Continue reading the next section.

7.1.1 Do you need a Certification?

Obtaining a certification in your specialty area can help you to build a solid career. Having a Professional Certification could give you a "Competitive Edge": the advantage of getting a job or promotion first!

A certification shows your employer (or a future employer) that you are committed to your career; that you take it seriously. There are some employers who hire *certified technicians* only. Some certifications can make a difference in your starting salary, just by having it!

Figure 98   CompTIA A+ Certification.

Simply explained, in a highly competitive and rapidly changing world, getting a certification is an *investment* for the future of the candidate. There are different certifications available, depending on the field in which you choose to work. In the case of Biomedical Engineering it is no exception. But what certifications are available for those working in the biomedical field?

7.1.2 CBET Certification

CBET stands for *"Certified Biomedical Equipment Technician"*. As mentioned on page 43, this certification is offered by AAMI and the ICC and is the official certification for Biomedical Technicians in the United States.

According to the ICC, the technician or engineer that obtains this certification *"have demonstrated excellence in theoretical as well as practical knowledge of the principles of biomedical equipment technology"*. **[20]**.

The exam consists of 150 questions, divided into the following areas: **[20]**

Subject Area	% of the Test
Anatomy & Physiology	13%
Public (employee, patient, visitor) Safety in the Healthcare Facility	17%
Fundamentals of Electricity, Electronics, and Solid-State Devices	17%
Medical Equipment Function & Operation	26%
Medical Equipment Problem Solving	27%

You have four hours to complete the test. The results take several weeks to arrive via regular mail.

If you are interested in taking the exam for this certification, AAMI and ICC provide study materials from books to CD-Roms. Visit AAMI (www.aami.org) for all the details, terms and conditions and test schedule.

Figure 99   The CBET Certification ("Certified Biomedical Equipment Technician") is the official Biomed Certification in the US.

7.1.3 CET- Biomedical Electronics (ETA)

ETA (Electronics Technicians Association) provides an alternative to the CBET exam. Their CET (Certified Electronics Technician) Certification has an option called "Journeyman", that is offered to technicians with two or more years of experience in the field. There is a specialty available known as "Biomedical Electronics" or "BMD".

Figure 100   CET ("Certified Electronics Technician") with the "BMD" (Biomedical) Specialty.

I mentioned that this is an "alternative" because, to obtain this certification, ETA offers a fairly extensive list of 22 topics that cover almost the same topics of the CBET exam. You will need to study diligently, but it will help prepare you for the CBET exam.

On the ETA website there is a PDF document with a list of all the topics you need to study and includes a suggested reading list to prepare for the test. You can find the document at this address:
http://www.eta-i.org/Comps/BMD Comps.pdf .

7.2 Associations, Trade Shows and Biomed Conferences

There are several Biomed Associations and other organizations with local chapters in many areas. These associations are usually created by technicians, biomedical engineers and other professionals who wish to share information, experiences and knowledge with the idea of staying "up-to-date" with technological innovations and the changes occurring in the medical field.

In these Associations, vendors and technicians from various companies meet and discuss information of mutual interest. Also, by sharing methodologies and repair processes, new ideas are develop to offer better quality service.

Some of these associations provide tutorials on how to get your CBET license and offer discounts to buy study materials. Most Associations have their own website and sometimes they offer their own publications (newsletters) where job vacancies, events and seminars are advertised. This is important because attending technical seminars helps you to acquire new knowledge, and also helps you get "Continuing Education" points necessary to maintain your license as a Biomedical Technician.

Figure 101   BAAMI Website (Bay Area Association of Medical Instrumentation in Tampa Bay, FL). (www.Baami.org)

Biomedical Associations have regular meetings and sponsor large conventions annually, known as "Trade Shows". They invite many companies and manufacturers to showcase their new technologies, tools and repair processes. It is an excellent opportunity to compare equipment from different manufacturers, costs and benefits.

Figure 102 Visiting "Trade Shows" will help you to learn about Biomed Manufacturers and new technologies.

It is important that while visiting the "Trade Shows" booths, ask for "business cards" of the vendors and other printed material from the manufacturers. Sometimes they provide CD ROM's to share technical information or their most recent catalog. This information can be useful for you in the future.

Figure 103 Electronics Equipment convention. (Courtesy of Alpha Electronics - www.alphaelectronics.net)

Some manufacturers even have job applications forms available during a trade show! So I'll suggest something very important: bring copies of your current resumé when visiting one of these conventions in the future. You could have a pleasant surprise!

On the other hand, I've attended meetings offered by different Biomedical Associations. In those meetings, representatives from different Biomed companies are invited to offer a presentation, free of charge. This is an advantage for both the technicians that attend and for the company that provides the seminar. Why? The technicians will acquire new knowledge, while the company will advertize their product or service! So it's a 'win-win" situation for everybody involved.

7.2.1 Networking

One of the benefits of being a member of a Biomed Association is that you can do *"networking"*: having the opportunity to meet many people that work in the same technology field, from technicians to vendors. Having these contacts will help you immensely in your career, so get as many references as possible!

Figure 104   Being a member of a Biomed Association will help you meet many people that work in the field. (Cortesy of Baycare Clinical Engineering Services).

In fact, author Sharon Hanna in her book *"Career by Design"* **[23]** specifies how useful is to acquire references of this kind. She mentions (among other things) that, even before you start to look for a job, *start to meet people in the field*. Ask for their business cards, or at least their names, phone numbers and email address. One of the main ways to achieve this, the author explained, is *joining a professional organization*! How can you use all this information?

By meeting so many people in different areas of the biomedical field, you can learn about what kind of work your future peers do, the salary paid for a similar position and your colleagues can refer you to employment opportunities. You can use peers that know you as references for a job interview. And you can share information or clarify your doubts or questions with people who have more experience.

So, as a student, familiarize yourself with your local Biomed Association. Some have a very affordable "student membership" and others allow you to attend their meetings for free if you have a university/Student ID. Search the Internet or your local "Yellow Pages" to contact them.

7.3 Professional publications and the Internet

There is no doubt that the Internet has been the best invention ever created to share information. Never in human history has there been a "library" so vast, with the information immediately accessible around the world!

The Internet is a great tool, especially when it is used the right way. How can we use the Internet for the benefit of our Biomed career?

Think about it: At this moment as a student, you have a huge flow of information at your fingertips, without having to leave home or go to the library! However, since not all the information you find online is reliable, it is a good idea to always investigate the source. Any article or study that you find online should have a bibliography that indicates the information sources used to write them.

I suggest that, from time to time, visit the websites of biomedical equipment manufacturers. That will give you an idea how many manufacturers exist and the variety of biomed equipment available. I'll give you some suggestions *****:

- www.gehealthcare.com
- www.medical.siemens.com
- www.steris.com
- www.healthcare.philips.com
- www.cardinal.com
- www.datascope.com
- www.welchallyn.com
- www.nellcor.com

- www.puritanbennett.com
- www.valleylab.com
- www.masimo.com

*(Note: This list of companies is only a suggestion. It does not mean that the author endorse them or has interests in these companies).

What other online tools are available?

On the Internet there are <u>many</u> online encyclopedias that can help you increase your knowledge of Biomedical and other equipment. These are some of the most popular:

- britannica.com
- wikipedia.org <u>*</u>

<u>*</u> (Note: Wikipedia is considered "open content": anyone can edit the information presented. Many teachers *do not* allow the use of Wikipedia as a Bibliographic reference. Use references to <u>reliable</u> sources available at the end of the articles.)

Besides online encyclopedias, there is a number of "search engines" that have excellent tools to search for information, photos, video, etc. Some even have translation tools!

Some well-known search engines are:

- www.google.com (it has a translation system for several languages in http://translate.google.com)
- www.altavista.com (has a translation system called "Babelfish")
- www.yahoo.com
- www.webcrawler.com (checks different "Search Engines" at the same time)
- www.lycos.com

As I mentioned above, something you should start to seek, are local biomed associations. Start your search for them in your favorite search engine!

There are also several professional magazines devoted exclusively to the Biomedical Field that can be very useful:

• www.24x7mag.com - Recommended! (You can subscribe to the printed magazine for **FREE!**)

• www.medicalimagingmag.com

www.aami.org (You must visit it regularly!)

www.ecri.org (Emergency Care Research Institute - provides alerts about biomedical equipment)

Figure 105   24X7 Magazine

• www.medgadget.com - Great place to find new medical devices.

As time passes, you'll be so familiar with the Biomedical field that it will be easy and natural for you to search and locate the information you need.

Conclusion

As a Biomedical Engineering student, you have entered into a gigantic world, with a lot of options, opportunities and chances for growth. What can I recommend to you?

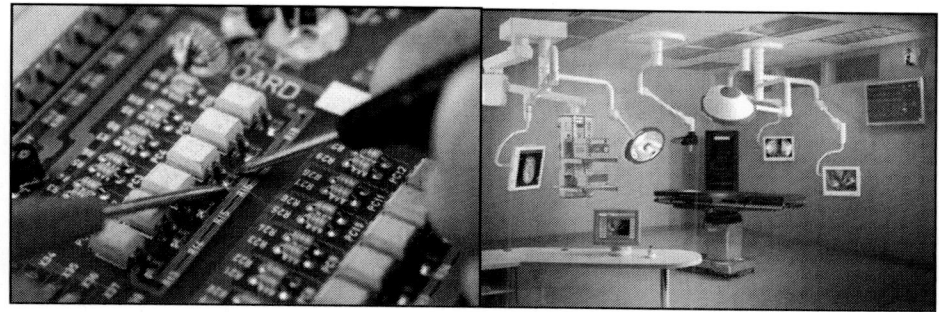

Courtesy of iStockPhoto.com and Steris.

- *Learn about your job at every opportunity.* As a student, nobody sees you as a "threat" to their job. It is the best time to expand your knowledge and benefit from the experience of other colleagues who have more experience in this field. Ask questions!

- *Never take "shortcuts":* Learn the proper way to do your job from the beginning. Remember that you will repair equipment that *people's lives depend on*! Follow **all** the steps and manufacturer's recommendations for the repair, for the preventive maintenance (PM), or to adjust or calibrate a medical unit.

- *Comply with all the law requirements* and suggestions from the regulatory agencies.

- *Be honest in all your dealings.* A good reputation will open many doors for you.

- *Share your experience.* We live in the "Information Age" *. If we share our knowledge with other colleagues, they will do the same for us.

Like everything in life, developing your skills involves time, effort and dedication. But it's worth it!

Good luck and success in all your future plans!

* Information Age - the current stage in societal development which began to emerge at the end of the twentieth century. This period is marked by the increased production, transmission, consumption of and reliance on information [38]

If you have suggestions, questions or comments about this publication, don't hesitate to contact the author visiting his website:

www.biomedtechnicians.com

Bibliography:

[1] University of Puerto Rico. *Plan para el Manejo de Desperdicios Médicos Regulados*. [Brochure]. Humacao, PR: Oficina de Salud, Seguridad ocupacional y Protección Ambiental. (http://www.uprh.edu/ssocupacional/pdf_doc/plan_mdbr.pdf)

[2] *Biomedical Engineer*. BMES - Biomedical Engineering Society. Retrieved August 24, 2007, from Biomedical Engineering Society Web site: http://www.bmes.org/careers.asp

[3] Enderle, J., Bronzino, J., & Blanchard, S (2005). *Introduction to Biomedical Engineering*. Elsevier Science & Technology Books, Page 20.

[4] *Glosario de Términos Cardíacos*. Retrieved November 10, 2007, from Medical Center Hospital Web site: http://www.mchodessa.com/spanish/PatientCare/CardiacTerms.htm

[5] Grant, Mark. *Issues with Leakage*. Retrieved November 12, 2007, from Marc's Technical Pages: Power Quality Symptoms and Solutions Web site: http://www.marcspages.co.uk/pq/3220.htm

[6] *Fire Prevention and Safety During Surgical Procedures*-Glossary. Retrieved September 16, 2007, from Covidien Energy-Based Professional Education Web site: https://www.valleylabeducation.org/fire/pages/fire-glossary.html

[7] Utz, Jeffrey. *What percentage of the human body is composed of water?* MadSci Network, Retrieved January 13, 2008, from http://www.madsci.org/posts/archives/2000-05/958588306.An.r.html

[8] Waxman, Henry A. (1990, November 28). *To amend the Federal Food, Drug, and Cosmetic Act to make improvements in the regulation of medical devices, and for other purposes*. Retrieved July 10, 2008, from The Library of Congress - Bills, Resolutions Web site: http://thomas.loc.gov/

[9] *Heart Electrical Phenomena*. Retrieved August 24, 2007, from hyperphysics.phy-astr.gsu.edu Web site: http://hyperphysics.phy-astr.gsu.edu/hbase/biology/heartelec.html

[10] Dorlands Medical Dictionary. Retrieved August 24, 2007, from mercksource.com Web site: http://www.mercksource.com.

[11] Leslie R. Atles, Scott Segalewitz (1995). *Affinity Reference Guide for Biomedical Technicians*. Marquette Electronics: Kendall/Hunt Publishing Co. *Page 7*.

[12] *Sphygmomanometer* - Definition. In *Merriam-Webster Online Dictionary* [Web]. Springfield, MA : Merriam-Webster, Inc.. Retrieved 01/15/2008, from http://www.merriam-webster.com/dictionary/sphygmomanometer .

[13] Defibrillator LifePoint Bag Manufacturer. Retrieved March 7, 2008, from Alibaba Global Trade Web site: http://www.alibaba.com/catalog/11538756/LifePoint_Bag.html

[14] Brain, Marshall. *Implantable Cardioverter-defibrillator*. Retrieved December 5, 2007, from howstuffworks.com Web site: http://healthguide.howstuffworks.com/implantable-cardioverter-defibrillator-dictionary.htm

[15] *Joules* - Glossary of Medical Terms. Retrieved Feb 9, 2008, from powermedix.com Web site: http://www.powermedix.com/glossary

[16] *Endo* - Glossary of HIV/AIDS Terms. Retrieved March 22, 2008, from The San Francisco AIDS Foundation Web site: http://www.sfaf.org/custom/glossary.aspx?l=en&a=E

[17] Sterilization. *The American Heritage® Science Dictionary*. Retrieved March 28, 2008, from Dictionary.com website: http://dictionary.reference.com/browse/telemetry.

[18] Telemetry. (n.d.). *The American Heritage® Science Dictionary*. Retrieved March 28, 2008, from Dictionary.com website: http://dictionary.reference.com/browse/telemetry

[19] *CompTIA Certifications*. Retrieved January 26, 2008, from CompTIA- The Computing Technology Industry Association Web site: http://certification.comptia.org/candidates/default.aspx

[20] *AAMI: ICC/USCC Certification*. Retrieved April 14, 2008, from AAMI- Association for the Advancement of Medical Instrumentation Web site: http://www.aami.org/certification/faq.html

[21] Carr, J., & Brown, J. (1981). *Introduction to Biomedical Equipment Technology*. New York: John Wiley & Sons. Page 313.

[22] Panichello, J. (2005). *X-Ray Repair: A Comprehensive Guide to the Installation and Servicing of Radiographic Equipment*. Illinois: Charles C. Thomas Publisher, LTD. Page 5.

[23] Bronzino, J. (1995). *The Biomedical Engineering Handbook*. CRC Press, LLC/IEEE Press.

[24] Aston, R. (1990). *Principles of Biomedical Instrumentation and Measurement*. New York: Merrill-Macmillan Publishing Company.

[25] Hanna, S (2004). *Career by Design: Communicating Your Way to Success (3rd Edition)*. Prentice Hall.

[26] *Inflamable*. Portalelectricos - RETIE - *Definiciones*. Retrieved June 3, 2008, from PortalElectricos.com Web site: http://www.portalelectricos.com/retie/cap1definicionesg.php

[27] Harrington, D. (1994, July). Electrosurgery Fact and Fiction. *Biomedical Instrumentation & Technology Magazine*, 331-333.

[28] Lara Hopley, Jo Van Schalkwyk. *The whole ECG - a really basic ECG primer*. Retrieved July 6, 2008, from Anaesthetist.com Web site: http://www.anaesthetist.com/icu/organs/heart/ecg/Findex.htm

[29] (2005). Tecnología: *Como funciona un MRI*. Retrieved July 8, 2008, from Fordham Radiology Web site: http://www.fordhamradiology.com/spanish/es_technology.htm

[30] *MRI* - Gutiérrez, Rodolfo (2008). *Consultor de Tecnología de Imágenes Diagnósticas*. Retrieved July 8, 2008, from Blogspot.com Web site: http://tecnologiadeimagenesdiagnosticas.blogspot.com

[31] *Phaco*. Eye Health Glossary - phacoemulsification (phaco). Retrieved July 14, 2008, from Eyecarefashion.com Web site: http://www.eyecarefashion.com/eye.health.glossary.0.html.0.html

[32] *FDA Centers & Offices*. Retrieved June 3, 2008, from Food and Drug Administration Web site: http://www.fda.gov/AboutFDA/CentersOffices/default.htm.

[33] *Clinical Laboratory Technologists and Technicians*. Retrieved May 12, 2009, from Bureau of Labor Statistics, *Occupational Outlook Handbook, 2008-09 Edition*, on the Internet at U.S. Department of Labor Web site: http://www.bls.gov/oco/ocos096.htm.

[34] *La misión de OSHA*. Retrieved August 10, 2008, from OSHA Web site: http://osha.gov/as/opa/spanish/mission-sp.html.

[35] *About NFPA*. Retrieved March 12, 2009, from NFPA - National Fire Protection Association Web site: http://www.nfpa.org.

[36] *NIST General Information*. Retrieved May 12, 2009, from National Institute of Standards and Technology Web site: http://www.nist.gov/public_affairs/labs2.htm.

[37] *HIPAA - General Information*. Retrieved May 12, 2009, from Centers for Medicare & Medicaid Services Web site: http://www.cms.hhs.gov/HIPAAGenInfo.

[38] *Information Age - Glossary*. Retrieved May 12, 2009, from Cyber.law.harvard.edu Web site: http://cyber.law.harvard.edu/readinessguide/glossary.html.

Also from the author:

Spanish Version ©2008

ISBN # 978-0-615-24158-6